Lecture Notes in Bioinformatics

Subseries of Lecture Notes in Computer Science

Trey Ideker   Vineet Bafna (Eds.)

Systems Biology and Computational Proteomics

Joint RECOMB 2006 Satellite Workshops
on Systems Biology and on Computational Proteomics
San Diego, CA, USA, December 1-3, 2006
Revised Selected Papers

 Springer

Series Editors

Sorin Istrail, Brown University, Providence, RI, USA
Pavel Pevzner, University of California, San Diego, CA, USA
Michael Waterman, University of Southern California, Los Angeles, CA, USA

Volume Editors

Trey Ideker
University of California
Department of Bioengineering
San Diego, CA 92093, USA
E-mail: tideker@ucsd.edu

Vineet Bafna
University of California
Computer Science and Engineering Dept.
San Diego, CA 92093, USA
E-mail: vbafna@cs.ucsd.edu

Library of Congress Control Number: 2007931338

CR Subject Classification (1998): F.2, G.3, E.1, H.2.8, J.3

LNCS Sublibrary: SL 8 – Bioinformatics

ISSN	0302-9743
ISBN-10	3-540-73059-1 Springer Berlin Heidelberg New York
ISBN-13	978-3-540-73059-0 Springer Berlin Heidelberg New York

Springer is a part of Springer Science+Business Media

springer.com

© Springer-Verlag Berlin Heidelberg 2007
Printed in Germany

Typesetting: Camera-ready by author, data conversion by Scientific Publishing Services, Chennai, India
Printed on acid-free paper   SPIN: 12076215   06/3180   5 4 3 2 1 0

Preface

The RECOMB Satellite Conferences on Systems Biology and Computational Proteomics were held December 1–3, 2006, at La Jolla, California. The Systems Biology meeting brought researchers together on various aspects of systems biology, including integration of genome-wide microarray, proteomic, and metabolomic data, inference and comparison of biological networks, and model testing through design of experiments. Specific topics included:

- Pathway mapping and evolution in protein interaction networks
- Inference of protein signaling networks for understanding cellular responses and developmental programs
- Model prediction of drug mechanism of action and toxicity
- Multi-scale methods which bridge abstract and detailed models
- Systematic design of genome-scale experiments
- Modeling and recognition of regulatory elements
- Identification and modeling of *cis*-regulatory regions
- Modeling the structure and function of regulatory regions
- Comparative genomics of regulation

With the sequencing of the genome, and subsequent identification of the parts list (the gene and their protein products), there is a renewed emphasis on studying the proteome. This year, the computational proteomics meeting focused on on computational mass spectrometry. Mass spectrometry is emerging as a key technology for proteomics. The last few years have seen tremendous improvement in the quality and quantity of available peptide mass spectrometry data, as well as the realization that advanced computational approaches are critical to the success of this technology. The conference explored the use of this technology in various proteomic applications, including, but not limited to: protein identification and quantification in specific cellular environments; structural genomics; networks of protein interaction; post-translational modifications; and others.

We received approximately 50 full paper submissions to the joint workshops. After review, a total of 20 were invited for oral presentations, adding to 14 plenary talks. These papers appear either as extended abstracts in this volume or are published in the journal *Molecular Systems Biology*.

Finally, we gratefully acknowledge support from our sponsors: the International Society for Computational Biology, RECOMB Steering Committee, the California Institute for Telecommunications and Information Technology (Calit2), the UC Discovery Program, and Pfizer La Jolla.

December 2007

Vineet Bafna
Trey Ideker

Organization

Program Committee

Annette Adler	Agilent
John Aitchison	Institute of Systems Biology
Gary Bader	Memorial Sloan-Kettering Cancer Center
Vineet Bafna (Co-chair)	University of California, San Diego
Ron Beavis	University of British Columbia
Marshall Bern	Palo Alto Research Center
Tim Chen	University of Southern California
Eric Davidson	California Institute of Technology
Nathan Edwards	University of Maryland
Keith Elliston	Genstruct
Eleazar Eskin	University of California, San Diego
Tim Galitski	Institute of Systems Biology
Mark Gerstein	Yale University
Jeff Hasty	University of California, San Diego
Ralf Herwig	Max Planck Institute for Molecular Genetics
Leroy Hood	Institute for Systems Biology
Trey Ideker (Co-chair)	University of California, San Diego
Janette Jones	Unilever SEAC
Peter Karp	Bioinformatics Research Group, SRI Intl
Stuart Kim	Stanford University
Edda Klipp	Max Planck Institute for Molecular Genetics
Oliver Kohlbacher	Universität Tübingen
Douglas Lauffenburger	Massachusetts Institute of Technology
Mike Levine	University of California, Berkeley
Bin Ma	University of Western Ontario
Edward Marcotte	University of Texas
Andrew McCulloch	University of California, San Diego
Satoru Miyano	University of Tokyo
Alexey Nesvizhskii	University of Michigan
William Noble	University of Washington
Bernhard Palsson	University of California, San Diego
Dana Pe'er	Harvard Medical School
Pavel Pevzner	University of California, San Diego
Tzachi Pilpel	Weizmann Institute of Science
Teresa M. Przytycka	NIH/NLM/NCBI
Ben Raphael	Brown University
Knut Reinert	Freie Universität Berlin
Cenk Sahinalp	Simon Fraser University
Ron Shamir	Tel Aviv University

Roded Sharan	Tel Aviv University
Alfonso Valencia	Centro Nacional de Biotecnologia
Guy Warner	Unilever SEAC
Christopher Workman	Technical University of Denmark
John Yates	The Scripps Research Institute
Ralf Zimmer	Institut für Informatik

RECOMB Systems Biology Steering Committee

Trey Ideker (Chair)	University of California, San Diego
Ron Shamir	Tel Aviv University
Satoru Miyano	University of Tokyo
Douglas Lauffenburger	Massachusetts Institute of Technology
Leroy Hood	Institute for Systems Biology

RECOMB Computational Proteomics Steering Committee

Vineet Bafna (Chair)	University of California, San Diego
John Yates	The Scripps Research Institute
Tim Chen	University of Southern California
Pavel Pevzner	University of California, San Diego

Organizing Committee

Vineet Bafna	University of California, San Diego
Trey Ideker	University of California, San Diego
Nuno Bandeira	University of California, San Diego
Thomas Lemberger	MSB
BJ Morrison McKay	ISMB
Samantha Smeraglia	University of California, San Diego
Shaojie Zhang	University of California, San Diego

Sponsoring Institutions

The International Society for Computational Biology
Molecular Systems Biology
The UC Discovery Grant
Pfizer Inc.
The UCSD Jacobs School of Engineering
Calit2: California Institute for Telecommunications and Information Technology

Table of Contents

Not All Scale Free Networks Are Born Equal:
The Role of the Seed Graph in PPI Network Emulation

Fereydoun Hormozdiari[1], Petra Berenbrink[1], Nataša Pržulj[2], and Cenk Sahinalp[1]

[1] School of Computing Science, Simon Fraser University, Canada
[2] Department of Computer Science, University of California, Irvine, USA

Abstract. The (asymptotic) degree distributions of the best known "scale free" network models are all similar and are independent of the seed graph used. Hence it has been tempting to assume that networks generated by these models are similar in general. In this paper we observe that several key topological features of such networks depend heavily on the specific model and the seed graph used. Furthermore, we show that starting with the "right" seed graph, the *duplication model* captures many topological features of publicly available PPI networks very well.

1 Introduction

In the past few years protein-protein interaction (PPI) networks of several organisms have been derived and made publicly available. Some of these networks have interesting topological properties; e.g. the degree distribution of the Yeast PPI network is heavy tailed (i.e. there are a few nodes with many connections). It has been argued that the degree distribution of these networks are in the form of a *power-law*[14], [24].[1] Since well known random graph models also have power-law degree distributions [3], [8], [25] it has been tempting to investigate whether these models agree with other topological features of the PPI networks.

There are two well known models that provide power law degree distributions (see [10], [9], [4]). The *preferential attachment* model [2], [8], was introduced to emulate the growth of naturally occurring networks such as the web graph; unfortunately, it is not biologically well motivated for modeling PPI networks. The *duplication model* on the other hand [7], [22], [18] is inspired by Ohno's hypothesis on genome growth by duplication. Both models are iterative in the sense that they start with a *seed graph* and grow the network in a sequence of steps.

The degree distribution is commonly used to test whether two given networks are similar or not. However, networks with identical degree distributions can have very different topologies.[2] Furthermore, it was observed in [23] that given two networks with substantially different initial degree distributions, a partial (random) sample from

---

[1] Some recent work challenge this by attributing the power law like behavior to sampling issues, experimental errors or statistical mistakes [23], [16], [21], [19], [12].

[2] Consider, for example, an infinite two dimensional grid vs a collection of cliques of size 5; in both cases all nodes have degree 4.

T. Ideker and V. Bafna (Eds.): Syst. Biol. and Comput. Proteomics Ws, LNBI 4532, pp. 1–13, 2007.

those networks may give subnetworks with very similar degree distributions. Thus the degree distribution can not be used as a sole measure of topological similarity.

In the recent literature two additional measures have been used to compare PPI networks with random network models. The first such measure is based on the k-*hop reachability*. The 1-hop reachability of a node is simply its degree (i.e. the number of its neighbors). The k-hop reachability of a node is the number of distinct nodes it can reach via a path of $\le k$ edges. The k-hop reachability of all nodes whose degree is ℓ is the average k-hop reachability of these nodes. Thus the k-hop reachability (for $k = 2, 3, \ldots$) of nodes as a function of their degree can be a used to compare network topologies. An earlier comparison of the k-hop reachability of the Yeast network with networks generated by certain duplication models concluded that the two network topologies are quite different [5]. The second similarity measure is based on the *graphlet distribution*. Graphlets are small subgraphs such as triangles, stars or cliques. In [16] it was noted that certain "scale free" networks are quite different from the Yeast PPI network with respect to the *graphlet distribution*. This observation, in combination with that on the k-hop degree distribution seem to suggest that the known PPI networks may not be scale free and existing scale free network models may not capture the topological properties of the PPI networks.

There are other topological measures that have been commonly employed in comparing social networks etc. but not PPI networks. Two well known examples are the *betweenness* distribution and the *closeness* distribution [26]. Betweenness of a vertex v is the number of shortest paths between any pair of vertices u and w that pass through v, normalized by the total number of such paths. Closeness of v is the inverse of the total distance of v to all other vertices u. Thus one can use betweenness and the closeness distributions, which respectively depict the number of vertices within a certain range of betweenness and closeness values can be used to compare network topologies.

2 Network Generation Models

The two network network models we study here both start with a small seed graph and add one node to it in each iteration. Let $G(t) = (V(t), E(t))$ be the graph at the end of time step t, where $V(t)$ is the set of nodes and $E(t)$ is the set of edges/connections. Let v_t be the node generated in time step t. Given a node v_τ, we denote its degree at the end of time step t by $d_t(v_\tau)$.

Preferential attachment model. The preferential attachment model was analyzed in [2], [6], [8] ,[10]. In step t it generates v_t and connects it to every other node v_τ independently with probability $c \cdot d_{t-1}(v_\tau)/2|E(t-1)|$, where c is the average degree of a node in G; i.e. v_t prefers to connect itself to high degree nodes.

Duplication model. This model is based on Ohno's hypothesis of genome evolution [7], [18], [22]. In iteration t, a node v_τ of $G(t-1)$ is picked uniformly at random and "duplicated", i.e. an exact copy of v_τ as v_t is generated. The model then updates v_t's edges, first by deleting each of its edges with probability $(1-p)$, then by connecting each node $v_{t'}$ (except the neighbors of v_τ) to v_t independently with probability $r/|V(t)|$. Here, p and r are user defined parameters. Much of the earlier work on the duplication model aim to maintain a constant average degree throughout the generation of the network; this is achieved by setting $r = (1/2 - p).a$.

As mentioned earlier, the degree distribution of the preferential attachment model as well as the duplication model asymptotically approaches a power law [2], [8], [10], [9]. More specifically, in the log-log scale, it forms a straight line (this is valid for only "high degree" nodes) whose slope is independent of the seed graph and a function of the values of p and r for the duplication model or c for the preferential attachment model. Thus, the two iterative models are equivalent with respect to the degree distribution.

Both the preferential attachment and the duplication model produce many *singletons* [3] [4]. Singletons are nodes which are not connected to any other node. Unfortunately there are no known bounds on the number of generated singletons in the duplication model. In the duplication model, for the special case $r = 0, p = 1/2$, the proportion of singletons asymptotically approaches 1. However, the number of singletons in known PPI networks is very small.

Modified duplication model. It is well known that the number of singletons in PPI networks are quite limited. This does not come as a surprise as genes with no functionality are not conserved during evolution. Thus a slightly modified duplication model which deletes each singleton node as soon as it is generated may better emulate the growth of PPI networks. This model has also been shown to achieve a power law degree distribution [4].

Unfortunately, similar to the number of singletons in duplication model, in modified model the total number of generated nodes is not known. Moreover, it is not known which values of p and r ensure that the expected average degree is constant through all iterations. In Section 2.1 we derive conditions on p and r that are necessary for having a constant expected degree. We later use the derived relationship between p and r so that the modified duplication model can well approximate the desired average degree as well as the degree distribution of the PPI networks under investigation.

2.1 The Parameters of the Modified Duplication Model

Here we show how to determine conditions on deletion probability $1 - p$ and insertion probability r so that the expected average degree of the network can be set to any given value. For this, we make the the assumption that the degree frequency distribution and the average degree of nodes are fixed asymptotically once the values of p and r are determined. Let $G(t) = (V(t), E(t))$ be the network generated by the modified duplication model and let $n(t) = |V(t)|$ and $e(t) = |E(t)|$. Also, let $n_k(t)$ be the number of nodes in time step t with degree k and $a(t)$ be the average degree of nodes in $G(t)$. Finally let $P_k(t) = n_k(t)/n(t)$, the frequency of nodes with degree k at time step t. We assume that $P_t(k)$ is asymptotically stable, i.e. $P_k(t) = P_k(t + 1)$ for all $1 \le k \le t$ for sufficiently large values of t. In other words we assume that $P_k(t) = d_k$

---

[3] We also note that the known PPI networks have several self loops. Both the preferential attachment and the duplication models can be modified slightly to produce such self loops(homodimers).

for some fixed d_k. By definition

$$a(t) = \sum_{k=1}^{t} k \cdot \frac{n_k(t)}{n(t)} = \sum_{k=1}^{t} k \cdot P_k(t) = \sum_{k=1}^{t} k \cdot d_k.$$

Now we can calculate the average degree $a(t + 1)$ under the condition that degree frequency distribution is stable and $a(t) = a$, a constant.

$$Exp[e(t + 1)] = e(t) + \sum_{k=1}^{t} k \cdot P_k(t) \cdot p + r = \frac{n(t) \cdot a(t)}{2} + p \cdot a(t) + r.$$

Let $Pr_s(t)$ be the probability that v_{t+1} ends up as a singleton.

$$Pr_s(t) = \sum_{k=1}^{t} P_k(t) \cdot (1 - p)^k \cdot \left(1 - \frac{r}{n(t)}\right)^{n(t)-k} \approx \sum_{k=1}^{t} d_k \cdot (1 - p)^k \cdot \frac{1}{e^r}.$$

Since this probability does not depend on t asymptotically, we can set $Pr_s(t) = Pr_s$. Now we can calculate the expected number of nodes and the expected number of edges in step $t + 1$.

$$Exp[n(t + 1)] = Pr_s \cdot n(t) + (1 - Pr_s) \cdot (n(t) + 1).$$

$$Exp[e(t + 1)] = Exp\left[\frac{n(t + 1) \cdot a(t + 1)}{2}\right] = \frac{a}{2} \cdot Exp[n(t + 1)]$$

$$Exp[e(t + 1)] = \frac{a}{2} \cdot (Pr_s \cdot n(t) + (1 - Pr_s) \cdot (n(t) + 1)).$$

Comparing the above equation with the first equation for $Exp[e(t + 1)]$ we get

$$\frac{a}{2} \cdot (Pr_s \cdot n(t) + (1 - Pr_s) \cdot (n(t) + 1)) = \frac{n(t) \cdot a(t)}{2} + p \cdot a(t) + r = \frac{n(t) \cdot a}{2} + p \cdot a + r.$$

Solving the above equation results in $a = 2r/(1 - Pr_s - 2p)$ where Pr_s is a function of p, r and d_k only.

The discussion above demonstrates that the two key parameters p and r of the (modified) duplication model are determined by the degree distribution (more specifically the slope of the degree distribution in the log-log scale) and the average degree of the PPI network we would like to emulate. Perhaps due to the strong evidence that the seed network does not have any effect on the asymptotic degree distribution [5], the role of the seed network (the only free parameter remaining) in determining other topological features of the duplication model has not been investigated.

3   Measures for Comparing Networks

There are several topological features that can be used to test whether two networks are similar or not, starting at very rigorous measures like isomorphism, to very relaxed characteristics like the degree distribution. In this paper we focus on five such properties, namely the *degree distribution*, the *k-hop reachability*, the *graphlet frequency*, the *betweenness distribution* and the *closeness distribution*.

Isomorphy. Two networks G and G' are called *isomorphic* if there exists a bijective mapping F from each node of G to a distinct node in G', such that two nodes v and w are connected in G if and only if $F(v)$ and $F(w)$ are connected. G and G' are called *approximately isomorphic* if by removing a "small" number of nodes and edges from G and G' they could be made isomorphic. Ideally, a random graph model that aims to emulate the growth of a PPI network should produce a network that is approximately isomorphic to the PPI network under investigation. Unfortunately there is no known polynomial algorithm for testing whether two networks are (exactly or approximately) isomorphic or not.

k-hop reachability. Let $V(i)$ denote the set of nodes in V whose degree is i. Given a node v, denote by $d(v, k)$ its k-hop degree, i.e., the number of distinct nodes it can reach in $\leq k$ hops. Now we define $f(i, k)$, the k-hop reachability of $V(i)$ as

$$f(i, k) = \frac{1}{|V(i)|} \sum_{w \in V, d(w)=i} d(w, k).$$

Thus $f(i, k)$ is the "average" number of distinct nodes a node with degree i can reach in k hops; e.g. $f(i, 1) = i$ by definition.

Graphlet frequency. The graphlet frequency was introduced in [16] to compare the topological structure of networks. A graphlet is a small connected and induced subgraph of a large graph, for example a small triangle or a small clique. The *graphlet count* of a given graphlet g with r nodes in a given graph $G = (V, E)$ is defined as the number of distinct subsets of V (with r nodes) whose induced subgraphs in G are isomorphic to g. In this paper we consider all 141 possible graphlets/subgraph topologies with $3, 4, 5, 6$ nodes. Additionally, we consider cliques of sizes $7, 8, 9, 10$. We enumerate these graphlets as shown in Figure 6.

Betweenness distribution. The betweenness of a fixed node of a network measures the extend to which a particular point lies 'between' point pairs in the network $G = (V, E)$. The formal definition of betweenness is as follows. Let $\sigma_{x,y}$ be the number of shortest path from $x \in V$ to $y \in V$ for all pairs $x, y \in V$. (Note that $\sigma_{x,y} = \sigma_{y,x}$ in undirected graphs). Let $\sigma_{x,y}(v)$ be the number of shortest path from $x \in V$ to $y \in V$ which go through node v. The betweenness $\text{Bet}(v)$ of node v is now defined as follows.

$$\text{Bet}(v) = \sum_{(i,j) \in V, i,j \neq v} \frac{\sigma_{i,j}(v)}{\sigma_{i,j}}.$$

Closeness. For all $x, y \in V$, we define $d_{x,y}$ as the length of the shortest path between x and y. The closeness of a node $v \in V$ is defined as

$$\text{Cls}(v) = \frac{|V| - 1}{\sum_{i \in V} d_{v,i}}.$$

4 Results and Discussion

As mentioned above, scale free network generators such as the preferential attachment model and the duplication model can have very similar degree distributions under appropriate choice of parameters. Moreover, the degree distribution of these models converge to a power law degree distribution whose shape is determined solely by the edge deletion and edge insertion probabilities and not by the initial "seed" graph [10]. Hence, it has been tempting to assume that networks generated by these models are similar in general and the effect of the seed graph in shaping the topologies of these networks has largely been ignored in recent literature.

Unfortunately two networks with very similar degree distributions may have very different topologies. For example, a network generated by the preferential attachment and another generated by the duplication model may have very different k-hop reachability, graphlet, betweenness and closeness distributions while having almost identical degree distributions (see Section A.1). Furthermore two networks generated by the same duplication model (and hence have very similar degree distributions) can differ substantially in terms of the above topological measures, if their seed networks are different (see Section A.2).

If the seed selection makes such a difference in shaping the topology of the generated network, is it possible to select the "right" seed network so that all interesting topological features of the PPI networks in question can be captured? We answer this question positively by demonstrating that carefully chosen seeds can result in a network that is very similar to PPI networks we considered in terms of all of the above distributions.

The PPI networks we tested include (the largest connected component of) the complete Yeast PPI network [20] with 4902 proteins and 17200 edges (as of Jul 2006). We also tested the more accurate but much smaller CORE Yeast network [11] and the lesser developed Worm network [20] (see Section A.3).

The seed graph we used for capturing the Yeast PPI network basically has two highly connected cliques of respectively 10 and 7 nodes. There are a few additional nodes sparsely connected to the cliques in a random fashion (the total number of nodes was 50). This ensured that the (normalized) degree distribution of the Yeast PPI network as well as its clique frequency distribution (which turns out to be an important determinant of the overall graphlet distribution) were similar to that of the seed graph.

There are two additional parameters associated with the duplication model: p, the edge maintenance probability and r, the edge insertion probability. These two parameters alone determine the (asymptotic) degree distribution and the average degree of the generated network. We chose $p = 0.365$ and $r = 0.12$ so that the degree distribution of the duplication model matches that of the Yeast PPI network (see Section 2.1 for the exact mathematical expressions for p and r).

We used the duplication model to generate 4 independent networks each with 4902 vertices. The resulting networks are compared to the Yeast PPI network in terms of the k-hop reachability, the graphlet, betweenness, and closeness distributions in Figure 1.

Under all these measures, the Yeast network is very similar to those produced by the duplication model. In fact the duplication model we consider here provides much

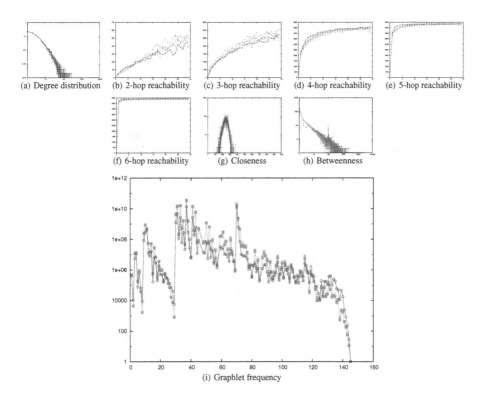

(a) Degree distribution (b) 2-hop reachability (c) 3-hop reachability (d) 4-hop reachability (e) 5-hop reachability

(f) 6-hop reachability (g) Closeness (h) Betweenness

(i) Graphlet frequency

Fig. 1. The degree distribution, the k-hop reachability, the graphlet, closeness and betweenness distributions of the Yeast PPI (Red) network against four independent runs of the duplication model (Green)

better fits to both the k-hop degree distribution and the graphlet distribution of the Yeast network than the random graph models described in of [5] and [16] - which were specifically devised to capture the respective features of PPI networks.

References

1. Alfarano, C., et al.: The biomolecular interaction network database and related tools. Nucl Acids Res. 33(Database Issue), 418–424 (2005)
2. Aiello, W., Chung, F., Lu, L.: Random graph model for power law graphs. In: Proc ACM STOC, pp. 171–180 (2000)
3. Barabási, A.-L., Albert, R.A.: Emergence of scaling in random networks. Science 286, 509–512 (1999)
4. Bebek, G., Berenbrink, P., Cooper, C., Friedetzky, T., Nadeau, J., Sahinalp, S.C.: The degree distribution of the general duplication models, Theor Comp Sci. (to appear)
5. Bebek, G., Berenbrink, P., Cooper,C., Friedetzky, T., Nadeau, J., Sahinalp, S.C.: Topological Properties of proteome networks. In: Proc RECOMB Sat. Mtg. on Sys. Bio. (LNBI) (2005)
6. Berger, N., Bollobás, B., Borgs, C., Chayes, J., Riordan, O.: Degree distribution of the FKP network model. In: Baeten, J.C.M., Lenstra, J.K., Parrow, J., Woeginger, G.J. (eds.) ICALP 2003. LNCS, vol. 2719, pp. 725–738. Springer, Heidelberg (2003)

7. Bhan, A., Galas, D.J., Dewey, T.G.: A duplication growth model of gene expression networks. Bioinformatics 18, 1486–1493 (2002)
8. Bollobás, B., Riordan, O., Spencer, J., Tusanády, G.: The degree sequence of a scale-free random graph process. Random Str. & Alg. 18, 279–290 (2001)
9. Chung, F., Lu, L., Dewey, G.T., Galas, J.D.: Duplication models for biological networks. J. Comp Bio. 10, 677–687 (2003)
10. Cooper, C., Frieze, A.: A general model of webgraphs. Random Str. & Alg. 22(3), 311–335 (2003)
11. Deane, C.M., Salwinski, L., Xenarios, I., Eisenberg, D.: Protein interactions: Two methods for assessment of the reliability of high-troughput observations. Mol. Cell Port 1, 349–356 (2002)
12. De Silva, E., Stumpf, M.P.H.: Complex networks and simple models in biology. J. of the Royal Society Interface 2, 419–430 (2005)
13. Erdös, P., Rényi, A.: On random graphs I. Publicationes Mathematicae Debrecen 6, 290–297 (1959)
14. Jeong, H., Mason, S., Barabasi, A.-L., Oltvai, Z.N.: Lethality and centrality in protein networks. Nature 411, 41 (2001)
15. Hermjakob, H., et al.: IntAct - an open source molecular interaction database. Nucl Acids Res. 32, 452–455 (2004)
16. Przulj, N., Corneil, D.G., Jurisica, I.: Modeling Interactome: Scale-Free or Geometric? Bioinformatics 150(1-3), 216–231 (2005)
17. Ohno, S.: Evolution by gene duplication. Springer, Heidelberg (1970)
18. Pastor-Satorras, R., Smith, E., Sole, R.V.: Evolving protein interaction networks through gene duplication. J. Theor Biol. 222, 199–210 (2003)
19. Przytycka, T., Yu, Y.K.: Scale-free networks versus evolutionary drift. Comp Bio. & Chem. 28, 257–264 (2004)
20. Salwinski, L., et al.: The Database of interacting Proteins:2004 update. Nucl Acids Res. 32(Database issue:D), 449–451 (2004)
21. Tanaka, R., et al.: Some protein interaction data do not exhibit power law statistics. FEBS Letters 579, 5140–5144 (2005)
22. Vázquez, A., Flammini, A., Maritan, A., Vespignani, A.: Modelling of protein interaction networks. Complexus 1, 38–44 (2003)
23. Han, J., Dupuy, D., Bertin, N., Cusick, M., Vidal, M.: Effect of sampling on topology predictions of protein-protein interaction networks. Nature Biotech 23, 839–844 (2005)
24. Wagner, A.: The Yeast protein interaction network evolves rapidly and contains few redundant duplicate genes. Mol. Biol Evol 18, 1283–1292 (2001)
25. Watts, D.J.: Small Worlds: The Dynamics of Networks between Order and Randomness. Princeton Univ Press, Princeton, NJ (1999)
26. Wasserman, S., Faust, K.: Social network analysis: methods and applications. Cambridge Univ Press, New York (1994)
27. Xenarios, I., et al.: DIP, the Database of Interacting Proteins: a research tool for studying cellular networks of protein interactions. Nucl Acids Res. 30, 303–305 (2002)
28. Zanzoni, A. et al.: MINT: a Molecular INteraction database. FEBS Letters 513(1), 135–140 (2002)

A Appendix

A.1 Duplication vs Preferential Attachment Models

In this section we show that the modified duplication model and the preferential attachment model with similar degree distributions may have very different k-reachability and graphlet distributions, thus considering only one of these models as a representative of "scale free" networks can be misleading.

Figure 2 depicts the degree distribution, k-hop reachability and graphlet frequency of the duplication model and the preferential attachment model with 4902 nodes (as per they Yeast PPI network [20]). We set $r = 0.12$, $p = 0.365$ and $d = 7$ so that the average degree of nodes in both models is 7 (again as per the Yeast PPI network [20]). Figure 2 compares the k-hop reachability achieved by the two models for $k > 1$. As can be seen, the k-hop reachability is quite different especially for $k = 3, 4$. Figure 2 also shows how the graphlet distributions differ, especially for dense graphlets (e.g graphlets 17-29 and 85-145). In terms of betweenness and closeness there are some differences as well.

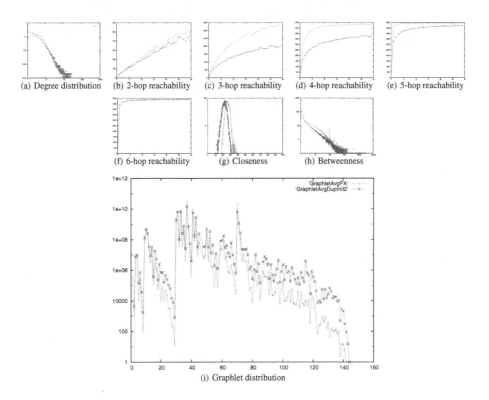

(a) Degree distribution   (b) 2-hop reachability   (c) 3-hop reachability   (d) 4-hop reachability   (e) 5-hop reachability

(f) 6-hop reachability   (g) Closeness   (h) Betweenness

(i) Graphlet distribution

Fig. 2. Degree distribution, k-hop reachability, graphlet, closeness and betweenness distributions of the preferential attachment model (Green) and the duplication model (Red)

A.2   The Effect of the Seed Network in Shaping the Topological Behavior of the Duplication Model

We now show that the seed network has a key role in characterizing the topology of the duplication model. Figure 3 depicts how various topological features of duplication models with identical parameters ($p = 0.365$ and $r = 0.12$) but different seed graphs vary. The first seed graph (red) is obtained by highly connecting two cliques of respective size 10 and 7 by several random edges. To reduce the average degree some additional nodes were generated and randomly connected to one of the cliques. The second seed graph (blue) is obtained by enriching a ring of 17 nodes by random connections so as to make the average degree match that of the first seed graph. The third seed graph (green) is formed by sparsely connecting two cliques of respective sizes 10 and 7 with some added nodes randomly connected to one of the cliques.

All three networks were grown until both had 4902 nodes as per the Yeast PPI network [20]. (We depict the average behavior of five independent runs of each of the models.) It can be observed that although all of them have very similar degree distributions, their graphlet distributions(Figure 3(i)) may be quite different, especially for dense graphlets. Note that the figure 3(i)and 3(g) are in logarithmic scale and seemingly small variations in the figure may imply several factors of magnitude of a difference between the two distributions. Figure 3 also compares the k-hop reachability, closeness and betweenness distributions. As can be seen the k-hop reachability and the closeness distribution can vary considerably.

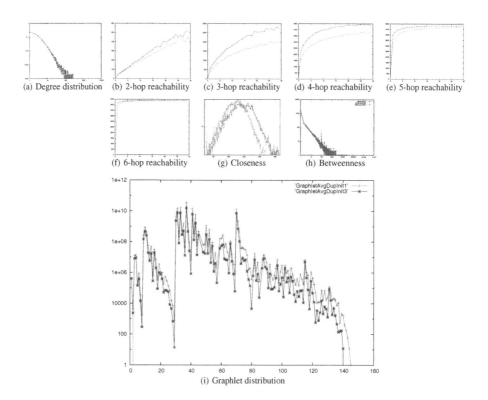

(a) Degree distribution   (b) 2-hop reachability   (c) 3-hop reachability   (d) 4-hop reachability   (e) 5-hop reachability

(f) 6-hop reachability   (g) Closeness   (h) Betweenness

(i) Graphlet distribution

Fig. 3. The effect of the seed network on the degree distribution, k-hop reachability, graphlet, closeness and betweenness distributions. Each color (Red, Blue, Green) depicts the behavior of a network with a particular seed graph. The parameters p and r are identical in all three models.

A.3   Duplication Model vs Other PPI Networks

We provide some additional evidence on the power of the duplication model in capturing the topological features of available PPI networks. We first compare the duplication model with the main component of the CORE subset of Yeast network. The CORE subset contains the pairs of interacting proteins identified in Yeast that were validated according to the criteria described in [11]. It involves 2345 nodes and 5609 edges. The values of r and p were set to $r = 0.12$, $p = 0.322$ as prescribed by the average degree formula $a = 2r/(1 - P_s - 2p)$ and the fact that P_s is a function of r and p. The seed network we used was very similar to that used for the complete Yeast network. The results are shown in Figure 4.

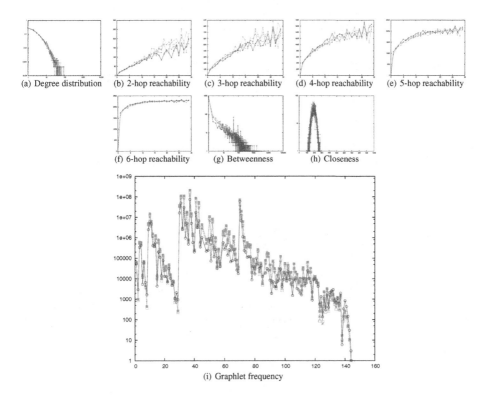

(a) Degree distribution   (b) 2-hop reachability   (c) 3-hop reachability   (d) 4-hop reachability   (e) 5-hop reachability

(f) 6-hop reachability   (g) Betweenness   (h) Closeness

(i) Graphlet frequency

Fig. 4. The topological properties of the duplication model (Green) compared to that of the CORE Yeast Network (Red). The degree distribution, the k-hop reachability, graphlet, betweenness and closeness distributions of both networks are shown. The values obtained by four independent runs of the duplication model are given.

We compare the duplication model with the Worm PPI network [20] as well. This network is much less developed than the Yeast network with only 2387 nodes and 3825 edges. The values of r and p we used for this network are $r = 0.12$, $p = 0.322$. The seed network we used was again very similar to that used for the Yeast network. The comparative results are shown in Figure 5.

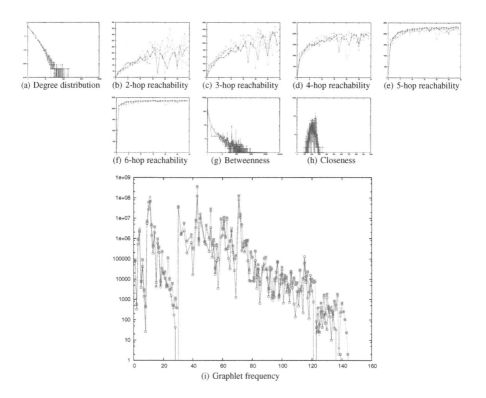

Fig. 5. The topological properties of the duplication model (Green) compared to that of the *C.Elegans* (Worm) network (Red). The degree distribution, the k-hop reachability, graphlet, betweenness and closeness distributions of both networks are shown. The results of four independent runs of the duplication model are depicted.

A.4 Graphlets

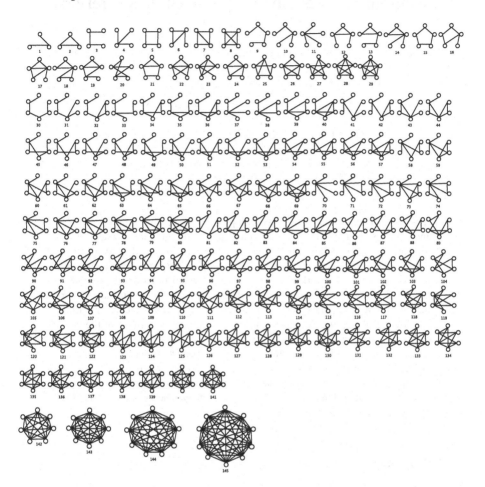

Fig. 6. The enumeration used for graphlet distributions

Probabilistic Paths for Protein Complex Inference

Hailiang Huang[1,2], Lan V. Zhang[3], Frederick P. Roth[3,4], and Joel S. Bader [1,2]

[1] Department of Biomedical Engineering, Johns Hopkins University, Baltimore, MD 21218
{hlhuang,joel.bader}@jhu.edu
[2] High-Throughput Biology Center, Johns Hopkins School of Medicine, Baltimore,
MD 21205
[3] Department of Biological Chemistry and Molecular Pharmacology,
Harvard Medical School, and
[4] Center for Cancer Systems Biology, Dana-Farber Cancer Institute, Boston, MA 02115
lanvzhang@gmail.com, fritz_roth@hms.harvard.edu

Abstract. Understanding how individual proteins are organized into complexes and pathways is a significant current challenge. We introduce new algorithms to infer protein complexes by combining seed proteins with a confidence-weighted network. Two new stochastic methods use averaging over a probabilistic ensemble of networks, and the new deterministic method provides a deterministic ranking of prospective complex members. We compare the performance of these algorithms with three existing algorithms. We test algorithm performance using three weighted graphs: a naïve Bayes estimate of the probability of a direct and stable protein-protein interaction; a logistic regression estimate of the probability of a direct or indirect interaction; and a decision tree estimate of whether two proteins exist within a common protein complex. The best-performing algorithms in these trials are the new stochastic methods. The deterministic algorithm is significantly faster, whereas the stochastic algorithms are less sensitive to the weighting scheme.

1 Introduction

The genome sequence of an organism provides a blueprint of its genes and proteins, but not the connections between these parts. Understanding how proteins are physically organized into complexes and pathways is increasingly based on observations from high-throughput experiments. Yeast has been the most widely used model for eukaryotic proteomics. High-throughput yeast two-hybrid screens have provided evidence for pair-wise links between proteins screens [1, 2]. Affinity purification followed by mass spectrometry identifies proteins that co-purify with a bait protein, suggesting shared membership in one or more protein complexes [3, 4].

Experimental interaction evidence can be unreliable due to high false-positive and false-negative rates [5, 6]. Experimental reports have included estimates of confidence based on multiple observations [1, 2, 7]. A more recent report of the fly protein interaction network included more sophisticated confidence metrics based on sequence analysis and network topology [8].

Here we consider three confidence-weighted networks derived from high-throughput data for yeast. The first, by Roth's group, is a naïve Bayes prediction (NB)

T. Ideker and V. Bafna (Eds.): Syst. Biol. and Comput. Proteomics Ws, LNBI 4532, pp. 14–28, 2007.
© Springer-Verlag Berlin Heidelberg 2007

of the posterior probability w_{ij} that two proteins have a direct physical interaction conditioned on observed data [9],

$$\frac{w_{ij}}{1-w_{ij}} = \left[\prod_\tau \frac{\Pr(x_{ij}^\tau \mid m_{ij})}{\Pr(x_{ij}^\tau \mid \overline{m}_{ij})} \right] \cdot \frac{\Pr(m)}{\Pr(\overline{m})}, \qquad (1)$$

where τ labels the different types of experimental data, x_{ij}^τ is the experimental data of type τ relating to protein pair i and j, m_{ij} indicates that the proteins have a direct physical interaction, $\overline{m}_{ij}$ indicates that the proteins do not have a direct physical interaction, and $\Pr(m)$ is the prior probability that two arbitrary proteins have a direct physical interaction.

The second network, by Bader and coworkers, predicts the probability of a direct or indirect physical interaction using a logistic regression model (LR) [10],

$$w_{ij} / (1 - w_{ij}) = \exp(\Sigma_\tau \beta_\tau x_{ij}^\tau + \Sigma_{\tau,\tau'} \beta_{\tau,\tau'} x_{ij}^\tau x_{ij}^{\tau'} + K) \cdot \Pr(m) / \Pr(\overline{m}). \qquad (2)$$

The model parameters $\{\beta\}$ were estimated using a training set equally weighted for true-positives and false-positives, equivalent to using 1 for the prior likelihood ratio $\Pr(m) / \Pr(\overline{m})$. Although the logistic regression scores have been used as the posterior probability of a true interaction [11], they are overconfident to the extent that non-interacting protein pairs outnumber interacting pairs in the true interaction network. A one-parameter fit for $\Pr(m) / \Pr(\overline{m})$ similar to that used for the NB network would convert the LR confidence scores to probabilities.

The final network, again by Roth's group, used a decision tree to estimate probabilities of protein pairs being co-complexed (DT) [12]. Then the odds of being co-complexed are multiplied by an adjustable parameter to estimate the odds of a direct physical interaction. This single parameter may then be fit to optimize performance for a training set. Unlike the NB model, the LR and DT models have the benefit of explicitly modeling dependence between predictors.

Other groups have used related methods to infer confidence-weighted edges not observed in the high-throughput data [13-20]. Some such methods include inference of shared complex membership or common function, training on just one complex or function at a time [21]. While we restricted our attention to the NB, LR, and DT weighting schemes, the methods we describe are directly applicable to other weighting schemes as well. Thus, the starting point for the methods we describe is an undirected weighted graph, in which proteins are represented as vertices and edge weights in the range [0,1] represent the probability of a direct or indirect physical interaction between proteins.

We investigate two general classes of algorithms that use confidence-weighted networks to infer protein complexes containing one or more seed proteins. First are deterministic algorithms, which directly calculate a threshold neighborhood around each seed protein, then identify proteins in the union of the neighborhoods as potential members of the complex. These algorithms include BESTPATH, published by Bader et al. as the SEEDY algorithm [22] and Shortest Path with Evidence (SPE), published previously by Roth's group [9] as a baseline for comparing improved algorithms. Here we report a new deterministic algorithm, SUMPATH, which attempts to combine information across multiple seeds.

The second class of algorithms generates a stochastic ensemble of networks using the edge weights as probabilistic measures that an edge taken from the high-throughput data is a true positive. This method was introduced by Roth's group in the PRONET algorithm [9], which requires the edge weights to refer to the probabilities of direct connections within a complex. Here we describe two related algorithms, PROPATH-ALG and PROPATH-EXP, designed to work well when edge weights also reflect the probability of indirect connections.

Although algorithms that are initialized with positive and negative seeds have been shown to be useful [21], the algorithms we describe use only positive seeds. Positive and negative seeds are particularly appropriate in the context of functional annotations using GO terms [23] or other ontologies in cases where terms in different lineages from the root are negatively correlated or mutually exclusive. The algorithms are also different from algorithms of finding complexes *de novo* [24, 25], which requires no seeds information.

Beyond introducing the new SUMPATH and PROPATH methods, the rationale of this report is to compare the abilities of each of these algorithms relative to recover well-annotated protein complexes when given partial information about these complexes. As in previous studies [9, 22], we use protein complexes from the MIPS catalog [26]. Furthermore, since the algorithms can be considered independently from the network confidence scores, we also compare performance as a function of the confidence score input. Because the PRONET algorithm was developed specifically for weights corresponding to direct connections, its performance is most fairly compared with other algorithms using the NB edge weights. Nevertheless, we provide results for all three networks using PRONET in the interests of completeness.

2   Methods

A summary of the algorithms is provided as Table 1. The input to each algorithm is a set of weighted edges $\{w_{ij}\}$ representing high-throughput interactions between proteins i and j, and a set of one or more seed proteins $\{s\}$. The output of each algorithm is a ranked list of other proteins in the network, where p_r is the protein with rank r in the list. Lower ranks correspond to greater probability that a protein is a member of a complex containing one or more of the seeds. In most of the algorithms, the ranks are calculated by first calculating a score S_i for each protein i, with higher scores corresponding to lower rank.

Each algorithm generates complex-membership scores differently based on the existence of one or more paths connecting seed proteins to other proteins in the network. For many proteins, no such path exists. These proteins are formally described as having distance = ∞ and/or score = 0 (the lowest possible value) and are appended to the end of the ranked list. We first describe the deterministic methods, Shortest Path with Evidence (SPE) [9], BESTPATH [22], and SUMPATH, then describe PRONET [9] and the probabilistic PROPATH algorithms.

Shortest Path with Evidence (SPE). The SPE method ignores the edge weights, treating each edge with any supporting evidence as having the same weight. The distance D_i of each protein in the network to the set of seeds is calculated as

$$D_i = \min_{s \in \text{seeds}} D_{is} \tag{3}$$

where D_{is} is the number of links in the shortest path connecting protein i to seed s, or $+\infty$ if no such path exists. Proteins are then ranked in decreasing order of D_i.

BESTPATH. The BESTPATH algorithm is identical to SEEDY, published earlier by Bader et al [22]. Here we term this algorithm BESTPATH to be more descriptive. With this algorithm, the weight of a path through proteins i_1, i_2, ..., i_n is the product of edge weights $\prod_{k=1}^{n-1} w_{i_k i_{k+1}}$. The score of each protein is defined as

$$S_i = \max_{s \in \text{seeds}} S_{is}, \tag{4}$$

where S_{is} is the highest weighted path between protein i and seed s. These paths may be computed efficiently using standard algorithms for traversing weighted graphs. Our implementation uses a priority queue implemented through a max-heap.

SUMPATH. We developed the SUMPATH method in an attempt to improve BESTPATH by searching for multiple high-weight paths. SUMPATH is based on Ising models for spin lattices [27, 28]. Each protein is assigned a spin label, 1 (part of the complex) or −1 (not part of the complex). Weighted edges in the network are interpreted as couplings between spins [29], and the goal is to identify the set of labels $\{S_i\}$ that minimize an energy function $-\Sigma_{(ij)} S_i w_{ij} S_j - \Sigma_i \phi_i S_i$, where ϕ_i is an external field representing prior knowledge of the probability of each spin state. Approximations such as mean field theory [28] or belief propagation [30] can be applied to reduce the computational complexity, but are beyond the scope of this paper. Here we present a simplified method. In this method, each seed s is assigned a score $S_s = 1$ that remains fixed throughout the algorithm. The BESTPATH method is used to initialize the scores $S_i^{(0)}$ of the other proteins. Scores for iteration $q + 1$ are obtained using the equations

$$T_i^{(q+1)} = \sum_j w_{ij} S_j^{(q)}, \quad Norm^{(q+1)} = \max_i T_i^{(q+1)} \quad \text{and} \quad S_j^{(q+1)} = T_i^{(q+1)} / Norm^{(q+1)} \tag{5}$$

to update the scores from iteration q. The sum over j in the first equation includes seed proteins. Iterations proceed until convergence, with 8-10 iterations required for convergence according to the criterion $\max_i \left| S_i^{(q+1)} - S_i^{(q)} \right| < 0.001$. The converged scores are then output. The normalization is required to prevent scores from growing without bound and is performed for the entire network rather than separately for each connected component.

PROPATH and PRONET Methods. PROPATH and PRONET are stochastic methods that require the generation of an ensemble of K replicate networks based on the edge weights. Each protein pair in each generated network receives a weight of either 0 or 1 based on a Bernoulli trial (i.e., a 'weighted coin flip') with probability w_{ij} that an edge between proteins i and j exists. Edges that are not included in the weighted network are assumed to have confidence 0 and never appear in a replicate network.

For each replicate network $k \in K$, the shortest path between protein i and seed s is denoted $D_{is}^{(k)}$, with $D_{is}^{(k)} = \infty$ if no path exists. These distances are calculated as with SPE, rather than BESTPATH, as the edge weights have already been taken into account in the generation of the replicate network. As with SPE, the distance to the closest seed is retained for each protein, $D_i^{(k)} = \min_s D_{is}^{(k)}$. If two proteins are in the same complex, we anticipate that multiple replicates in the ensemble will have a short path connecting the proteins. The mean distance over the ensemble, $K^{-1} \sum_k D_i^{(k)}$, is an inappropriate summary statistic because of the possibility that one of the replicates will generate an infinite distance.

The different PROPATH methods use distinct mathematical transforms to avoid this problem. Each transform maps infinite distance to zero score, and unit distance (the smallest possible distance for a protein that is not itself a seed) to unit score. The transforms we selected are

$$S_i^{(k)} = \begin{cases} I\left(D_i^{(k)} < \infty\right), & \text{PRONET} \\ \left(D_i^{(k)}\right)^{-\alpha}, & \text{PROPATH-ALG} \\ \exp\left(-\alpha D_i^{(k)} + \alpha\right), & \text{PROPATH-EXP} \end{cases} \tag{6}$$

where $S_i^{(k)}$ is the transformed score of protein i in replicate network k, $I(\text{arg})$ is an indicator function that is 1 for a true argument and 0 for a false argument, and α is a parameter defining the steepness of the decay of the algebraic or exponential transform. We have found that the PROPATH algorithms are insensitive to the exact value of α, with similar results for PROPATH-EXP for values of α up to 5 (a much faster decay; results not shown). For convenience, we used $\alpha = 1$ for PROPATH-ALG and PROPATH-EXP; performance may improve with an additional optimization over this single parameter.

In the PRONET algorithm, there is no distance-based decay. The existence of a path connecting a pair of vertices is converted to a 0/1 binary variable that is averaged over probabilistic networks. Formally, this is equivalent to taking the limit $\alpha \to 0$ in the PROPATH algorithms.

The final score of a protein is estimated as the average over replicates,

$$\hat{S}_i = K^{-1} \sum_k S_i^{(k)}. \tag{7}$$

The variance of $\hat{S}_i$ is bounded because $0 \le S_i^{(k)} \le 1$:

$$\mathrm{var}\left(\hat{S}_i\right) = K^{-1}\left(\left\langle S_i^2\right\rangle - \left\langle S_i\right\rangle^2\right) \le K^{-1}\left(\left\langle S_i\right\rangle - \left\langle S_i\right\rangle^2\right) \le \left(4K\right)^{-1}, \tag{8}$$

where the angle brackets refer to an average over a single replicate network. We used $K = 400$ to give a standard deviation no larger than 0.025. We checked that results had converged with respect to K.

Performance Metrics. We followed the same general procedure for each complex. First, we generated $N_{\mathrm{trial}} = 10$ random 50-50 splits of the complex into seed proteins and target proteins that were used as input to each algorithm. For complexes with an odd number of members, the seed group had one more member than the target group. The set of target proteins for trial t of complex c is denoted T_{ct}. The seeds were then used as input seeds for each of the algorithms, which returned lists of proteins ranked by decreasing likelihood of membership in the same complex as the seeds. Proteins used as seeds were omitted from the ranked list. The protein at rank r for trial t of complex c is denoted p_{ctr}. The indicator function $I\left(p_{ctr} \in T_{ct}\right)$ is 1 if this protein belongs to the target set and 0 otherwise.

Summing the indicator function over ranks, trials, and complexes provides a quantitative assessment of algorithm performance by generating a receiver operating characteristic (ROC) curve. The order of summation was as follows. First, for each complex and trial, we calculated the numbers of true positives and false positives through rank r, $TP_{ct}(r)$ and $FP_{ct}(r)$, as

$$TP_{ct}(r) = \sum_{r'=1}^{r} I\left(p_{ctr'} \in T_{ct}\right) \text{ and } FP_{ct}(r) = r - TP_{ct}(r). \tag{9}$$

This makes the conservative assumption that the identity of each complex is correctly reported in the MIPS data. The true positive and false positives counts were then averaged over the trials for each complex,

$$TP_c(r) = \left(N_{\mathrm{trial}}\right)^{-1}\sum_{t=1}^{N_{\mathrm{trial}}} TP_{ct}(r) \text{ and } FP_c(r) = r - TP_c(r). \tag{10}$$

The counts were then converted to true-positive and false-positive rates for each complex,

$$tp_c(r) = TP_c(r)/|T_c|, \quad fp_c(r) = FP_c(r)/|N_{\mathrm{tot}} - N_c|. \tag{11}$$

where $|T_c|$ is the cardinality of the target set for complex c, and $|N_{\mathrm{tot}} - N_c|$ is the number of proteins in the interaction network minus those that are also in the complex. Note that the maximum value for $tp_c(r)$ for large r is less than 1 if not every protein in the complex is in the protein interaction network. The maximum value of $fp_c(r)$ is 1, however. The overall true-positive and false-positive rates, averaged over complexes, are

$$tp(r) = C^{-1} \sum_{c=1}^{C} tp_c(r) \quad \text{and} \quad fp(r) = C^{-1} \sum_{c=1}^{C} fp_c(r). \tag{12}$$

This procedure gives equal weight to each complex. The ROC curve is the parametric plot of $tp(r)$ vs. $fp(r)$.

As with microarray analysis, the false-discovery rate may be more informative than the false-positive rate because the maximum number of false-positives far outweighs the maximum number of true-positives. The false-discovery rate is defined as a function of r as

$$fd(r) = C^{-1} \sum_{c=1}^{C} \frac{FP_c(r)}{TP_c(r) + FP_c(r)}. \tag{13}$$

With ~4000 proteins in the network, the false discoveries begin to dominate the returned list of proteins when the false-positive rate is on the order of N_{tot}^{-1}, or ~ 10^{-3}.

The area under the ROC curve (AUC) provides a quantitative measure of performance, with higher AUC corresponding to better performance. Our focus is on the region of the ROC curve with few false-positives. Thus, rather than calculating the area under the entire curve, we calculate the area up to a false-positive rate typical of what would be used in practice. We normalize this area to return a value termed $AUC(fp)$ that increases with better recall,

$$\text{AUC}(fp) = (fp)^{-1} \int_0^{fp} d(fp') tp(fp'), \tag{14}$$

where the true-positive rate is considered to be a function of the false-positive rate. Results are provided for AUC(0.1%) and AUC(0.5%). The AUC for a complex (AUCc) is also calculated to measure the complex specific recovery performance,

$$\text{AUC}_c(fp_c) = (fp_c)^{-1} \int_0^{fp_c} d(fp') tp_c(fp').$$

3   Results

Algorithms for extracting protein complexes from confidence-weighted interaction data were tested by assessing their ability to extract a known complex based on partial knowledge of its components. As a gold standard of true complexes, we used C = 23 known complexes from MIPS [26]. These complexes include many of those used in the original reports of the PRONET and BESTPATH algorithms. In general, each algorithm returns a ranked list of possible complex members and, based on the known complex, calculates recovery rates as a function of proteins through rank r : the true-positive rate $tp(r)$ (the fraction of positive predictions that are correct); the false-discovery rate $fd(r)$ (the fraction of positive predictions that are incorrect); and the false-positive rate $fp(r)$ (the fraction of non-interacting pairs that are predicted positive). Performance is visualized by graphing $tp(r)$ vs. $fp(r)$ in the region of $0 < fp(r) < 5 \times 10^{-3}$, corresponding to ~20 false-positives, and the $tp(r)$ vs. $fd(r)$ graph in the full region of $0 < fd(r) < 1$. Quantitative measures such as normalized AUC (Area Under the Curve) and FP-50 (false-positive rate at 50% recall) provide a convenient summary metric for ranking the algorithms (Table 1). The AUC for each complex (AUCc) is calculated (Fig. 2).

Table 1. Summary of methods. For each network, each algorithm was ranked 1-6 in performance, 1 = best, 6 = worst. Superscripts in numbers stand for the ranking, and are also indicated by the background colors (Green = rank 1 or 2; Yellow = rank 3 or 4; Red = rank 5 or 6; ties are colored as the best rank). The ranks were averaged to give an overall measure of each algorithm's performance. [a]Normalized area under the curve (AUC) at a false-positive rate of 0.1%, in percentage scale. See Eq. [14] for the normalization. [b]Normalized AUC at a false-positive rate of 0.5%. [c]False-positive rate at 50% recall, in percentage scale.

	Avg. Rank	AUC 0.1%(%)[a]			AUC 0.5%(%)[b]			FP-50(%)[c]			CPU Time (min)		
		NB	LR	DT	NB	LR	DT	NB	LR	DT	NB	LR	DT
PROPATH-EXP	2.25	10^1	17^1	20^1	19^1	34^1	34^1	7.3^1	1.2^2	1.0^2	8.1^2	240^5	2900^6
PROPATH-ALG	2.33	10^1	17^1	20^1	19^1	34^1	34^1	8.0^3	1.2^2	1.0^2	7.9^4	240^6	2700^5
BESTPATH	2.5	7.6^4	15^3	18^3	15^4	33^3	32^3	7.7^2	0.9^1	0.9^1	7.3^2	8.1^1	95^2
PRONET	3.8	10^1	0.03^5	18^3	19^1	0.18^5	31^4	9.5^5	27^6	1.6^4	7.6^3	230^4	2500^4
SPE	4.3	0.43^5	0.41^4	0.29^4	4.3^5	2.6^4	1.4^6	11^6	5.8^4	9.3^5	4.8^1	5.8^1	56^1
SUMPATH	4.8	1.3^5	0.009^6	4.1^6	12^5	0.092^6	8.4^5	8.1^4	22^5	5.8^4	77^6	42^3	300^3

NB network. We first compared algorithm performance for the confidence scores taken from NB [9] (Table 1 and Fig. 1A, B). The AUC (0.1%) and AUC (0.5%) measures show that PROPATH-EXP, PROPATH-ALG and PRONET have roughly equivalent performance in the region of stringent prediction, followed by BESTPATH. The SUMPATH algorithm has intermediate performance, and the SPE has the worst performance.

LR network. We then compared algorithm performance for confidence-weighted edges taken from LR [10]. The PROPATH-EXP and PROPATH-ALG algorithms perform the best and are comparable, followed closely by BESTPATH (Fig. 1C, D). These three algorithms dominate the other algorithms in this region of stringent prediction, returning ~40-50% of the target proteins.

DT network. The last edge weights we used are from DT [12]. This set of edge weights has a tunable parameter α. For each algorithm, we chose the value of α that maximized its AUC(0.5%). PROPATH-EXP and PROPATH-ALG have equivalent performance, followed closely by PRONET and BESTPATH. The remaining algorithms, SUMPATH and SPE, have the worst performance (Fig. 1E, F).

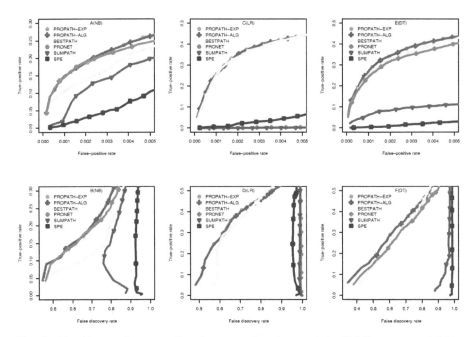

Fig. 1. Algorithm performance. Receiver operating characteristic (ROC) curves and false-discovery rates characterize the performance of algorithms to extract protein complexes from protein interaction networks. Fig. 1A and 1B are from edge weights using NB [9], Fig. 1C and 1D are from edge weights using LR [10] and Fig. 1E and 1F are from edge weights using DT [12].

Complex-specific recovery. We then investigated whether certain complexes are easier to recover than others. Given a set of network edges and a recovery algorithm, a one-sided Wilcoxon test was used to test the significance of the hypothesis that a particular complex had a higher than average AUC 0.5% compared to other complexes recovered using the same network edges and the same algorithm. A more complete description is provided in the Methods.

We found that the best-performing algorithms (PROPATH-EXP, PROPATH-ALG, and BESTPATH) consistently recovered four complexes with a higher than average AUC 0.5% regardless of the network edges used: the PROTEASOME, HISTONEAC, HISTONEDEAC and NUCLEARPORE. One reason for better-than-average recovery of these specific complexes may be the number of proteins contained in these gold-standard examples, 36, 17, 4, and 24 respectively. These are less than the mean number of proteins across all complexes, 45.8. A possible interpretation is that these four represent distinct single complexes. Other gold-standard complexes may in fact comprise a number of more loosely coupled sub-complexes that are more difficult to recover as single cohesive units. Such sub-complexes might also be expected to have more interactions outside the gold-standard complex, which would reduce the AUC. Furthermore, signaling pathways might also be expected to be more loosely coupled and not recovered as well.

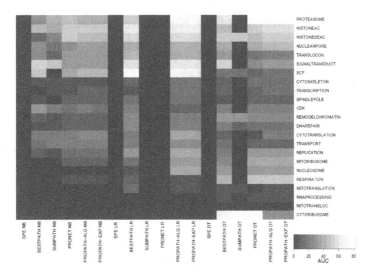

Fig. 2. Complex-specific performance plot identifies complexes that are better extracted using our methods. AUC 0.5% for each complex is visualized using the color key on the bottom right corner. Complexes have been reordered to show clusters of similar performance.Lighter colors indicate better performance.

Recovery performance may be visualized using a color-coded display of true-positives and nominal false-positives predicted by an algorithm (Fig. 3). We focus on a specific complex, histone acetyltransferase (HAC), which has 17 members. We generated random 50-50 splits of the complex into seed proteins (E_t) and target proteins. The seed proteins were then used as input to PROPATH-EXP with LR to generate a list of proteins ranked by decreasing likelihood of their memberships in HAC. We kept the first half of the ranked list, $P_t = \left\{ p_{rt}, r \leq \dfrac{N_{tot} - N_c/2}{2} \right\}$, excluding the seeds. N_{tot} is the total number of proteins in the list and N_c is the number of proteins in the complex. The number of times protein p has been used as seed is $N_{sp} = \sum_{t=1}^{N_{trial}} I\left(p \in E_t\right)$, where $I\left(p \in E_t\right)$ is the indicator function, $I\left(p \in E_t\right) = \begin{cases} 1 & p \in E_t \\ 0 & p \notin E_t \end{cases}$. The maximum possible recovery count for protein p is $N_{trial} - N_{sp}$, and the recovery rate for protein p is $R_p = \sum_{t=1}^{N_{trial}} I\left(p \in P_t\right) \Big/ \left(N_{trial} - N_{sp}\right)$. We defined three categories of recovered proteins:

$\{p \in \mathrm{HAC}\} \cap \{p, \text{with } R_p \geq 0.5\}$, High recovery rate true-positive protein

$\{p \in \mathrm{HAC}\} \cap \{p, \text{with } R_p < 0.5\}$, Low recovery rate true-positive protein

$\{p \notin \mathrm{HAC}\} \cap \{p, \text{with } R_p \geq 0.5\}$, High recovery rate false-positive protein

In the graph, we have 9 out of 17 HAC proteins recovered with $R \geq 0.5$ and 2 false positive proteins with $R \geq 0.5$. Despite not being included in the MIPS catalogue for HAC, these two proteins, SGF29 and SGF73, are annotated in SGD as probable subunits of the SAGA HAC.

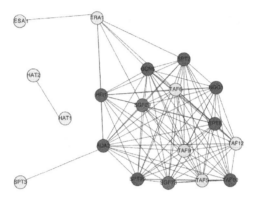

Fig. 3. Complex recovery graph. Histone acetyltransferase complex recovery graph shows the rate of proteins being recovered. Dark green nodes indicate high recovery rate true-positive proteins. Light green nodes indicate low recovery rate true-positive proteins. Red nodes indicate high recovery rate false-positive proteins.

Beyond recall performance, CPU performance may also be a criterion for selecting an algorithm. Timings are provided for a Perl implementation of each algorithm (FreeBSD 5.2.1, 3.0Ghz Pentium-4 CPU, 1GB memory). The deterministic algorithms SPE and BESTPATH are approximately 3 to 5 times faster than the probabilistic PROPATH algorithms. A naïve expectation is that the running time would scale as the number of probabilistic replicates sampled for the PROPATH algorithms; the difference is likely due to initialization overhead common to the probabilistic and deterministic algorithms. The SUMPATH algorithm, although deterministic, requires iterations for convergence. Thus, it is much slower than the other deterministic algorithms by about three times. The same algorithms implemented in C run an order of magnitude faster or more than those implemented in Perl, depending on the size of the network, but the relative timings of the algorithms are similar.

4   Discussion and Conclusion

We have introduced novel algorithms for predicting additional members of a protein complex based on knowledge of a subset of known members and access to a database of confidence-weighted protein-protein interactions. These algorithms have been tested against one another, and with related algorithms described previously in the literature. Important future work is to benchmark these algorithms against other methods that predict process-specific networks [31] or model the dynamical structure of protein complexes [32, 33]. Such comparisons will require standardized data sets and performance criteria [34].

The best-performing algorithms overall, PROPATH-EXP and PROPATH-ALG, share two distinctive characteristics. First, they rely on probabilistic sampling of protein interaction networks based on the confidence weights. Second, they use a distance measure, rather than the mere existence of a path, to rank potential complex members. A deterministic algorithm that performs almost as well in this test, BESTPATH, uses a greedy approach to identify the single path with greatest probability, but does not explicitly consider the length of a path. We attempted to improve the performance of BESTPATH by incorporating multiple paths. The resulting SUMPATH algorithm performed worse, however. The BESTPATH algorithm has an additional speed advantage over all other algorithms tested, excepting the poorly performing SPE method, which ignores confidence weights.

An important conclusion of this work is that algorithms may be sensitive to the meaning of an edge, in particular whether it represents a direct physical interaction or a more general functional association (such as co-membership in a protein complex). The PRONET algorithm, which was developed specifically for inference based on a network of direct interactions, indeed performs less well beyond its intended range. Other algorithms, including BESTPATH and PROPATH, appear more robust to the inclusion of indirect interaction edges. Quantitative measures of performance can depend on the examples used for testing; we find that some complexes are consistently recovered better than others regardless of algorithm or network edges.

While BESTPATH performed nearly as well as PROPATH-EXP and PROPATH-ALG in this test, we anticipate that the performance of BESTPATH will degrade in networks with many interaction edges having a weight close to 1, which should happen increasingly often as individual interactions are experimentally validated. As the number of high-weight edges increases, the BESTPATH algorithm will necessarily return an increasingly large fraction of proteins in the network. In this regime, however, the probabilistic PROPATH-EXP and PROPATH-ALG algorithms that explicitly consider the length of a high-confidence path should continue to give good performance.

Acknowledgements. JSB acknowledges support from NIH/NIGMS 1R01GM067761-01, NIH/NCRR U54RR020839, NSF 0546446, and the Whitaker Foundation. FPR acknowledges support from the Keck Foundation and NIH grants HG0017115, HG003224, & HL81341. LVZ was supported by Fu, Ryan, and AAUW Fellowships.

References

1. Uetz, P., Giot, L., Cagney, G., Mansfield, T.A., Judson, R.S., Knight, J.R., Lockshon, D., Narayan, V., Srinivasan, M., Pochart, P., Qureshi-Emili, A., Li, Y., Godwin, B., Conover, D., Kalbfleisch, T., Vijayadamodar, G., Yang, M., Johnston, M., Fields, S., Rothberg, J.M.: A comprehensive analysis of protein-protein interactions in Saccharomyces cerevisiae. Nature 403, 623–627 (2000)
2. Ito, T., Chiba, T., Ozawa, R., Yoshida, M., Hattori, M., Sakaki, Y.: A comprehensive two-hybrid analysis to explore the yeast protein interactome. Proc Natl Acad Sci. USA 98, 4569–4574 (2001)
3. Gavin, A.C., Bosche, M., Krause, R., Grandi, P., Marzioch, M., Bauer, A., Schultz, J., Rick, J.M., Michon, A.M., Cruciat, C.M., Remor, M., Hofert, C., Schelder, M., Brajenovic, M., Ruffner, H., Merino, A., Klein, K., Hudak, M., Dickson, D., Rudi, T., Gnau, V., Bauch, A., Bastuck, S., Huhse, B., Leutwein, C., Heurtier, M.A., Copley, R.R., Edelmann, A., Querfurth, E., Rybin, V., Drewes, G., Raida, M., Bouwmeester, T., Bork, P., Seraphin, B., Kuster, B., Neubauer, G., Superti-Furga, G.: Functional organization of the yeast proteome by systematic analysis of protein complexes. Nature 415, 141–147 (2002)
4. Ho, Y., Gruhler, A., Heilbut, A., Bader, G.D., Moore, L., Adams, S.L., Millar, A., Taylor, P., Bennett, K., Boutilier, K., Yang, L., Wolting, C., Donaldson, I., Schandorff, S., Shewnarane, J., Vo, M., Taggart, J., Goudreault, M., Muskat, B., Alfarano, C., Dewar, D., Lin, Z., Michalickova, K., Willems, A.R., Sassi, H., Nielsen, P.A., Rasmussen, K.J., Andersen, J.R., Johansen, L.E., Hansen, L.H., Jespersen, H., Podtelejnikov, A., Nielsen, E., Crawford, J., Poulsen, V., Sorensen, B.D., Matthiesen, J., Hendrickson, R.C., Gleeson, F., Pawson, T., Moran, M.F., Durocher, D., Mann, M., Hogue, C.W., Figeys, D., Tyers, M.: Systematic identification of protein complexes in Saccharomyces cerevisiae by mass spectrometry. Nature 415, 180–183 (2002)
5. Deane, C.M., Salwinski, L., Xenarios, I., Eisenberg, D.: Protein interactions: two methods for assessment of the reliability of high throughput observations. Mol. Cell Proteomics 1, 349–356 (2002)
6. von Mering, C., Krause, R., Snel, B., Cornell, M., Oliver, S.G., Fields, S., Bork, P.: Comparative assessment of large-scale data sets of protein-protein interactions. Nature 417, 399–403 (2002)
7. Li, S., Armstrong, C.M., Bertin, N., Ge, H., Milstein, S., Boxem, M., Vidalain, P.O., Han, J.D., Chesneau, A., Hao, T., Goldberg, D.S., Li, N., Martinez, M., Rual, J.F., Lamesch, P., Xu, L., Tewari, M., Wong, S.L., Zhang, L.V., Berriz, G.F., Jacotot, L., Vaglio, P., Reboul, J., Hirozane-Kishikawa, T., Li, Q., Gabel, H.W., Elewa, A., Baumgartner, B., Rose, D.J., Yu, H., Bosak, S., Sequerra, R., Fraser, A., Mango, S.E., Saxton, W.M., Strome, S., Van Den Heuvel, S., Piano, F., Vandenhaute, J., Sardet, C., Gerstein, M., Doucette-Stamm, L., Gunsalus, K.C., Harper, J.W., Cusick, M.E., Roth, F.P., Hill, D.E., Vidal, M.: A map of the interactome network of the metazoan C. elegans. Science 303, 540–543 (2004)
8. Giot, L., Bader, J.S., Brouwer, C., Chaudhuri, A., Kuang, B., Li, Y., Hao, Y.L., Ooi, C.E., Godwin, B., Vitols, E., Vijayadamodar, G., Pochart, P., Machineni, H., Welsh, M., Kong, Y., Zerhusen, B., Malcolm, R., Varrone, Z., Collis, A., Minto, M., Burgess, S., McDaniel, L., Stimpson, E., Spriggs, F., Williams, J., Neurath, K., Ioime, N., Agee, M., Voss, E., Furtak, K., Renzulli, R., Aanensen, N., Carrolla, S., Bickelhaupt, E., Lazovatsky, Y., DaSilva, A., Zhong, J., Stanyon, C.A., Finley, R.L., White Jr., K.P., Braverman, M., Jarvie, T., Gold, S., Leach, M., Knight, J., Shimkets, R.A., McKenna, M.P., Chant, J., Rothberg, J.M.: A Protein Interaction Map of Drosophila melanogaster. Science 302, 1727–1736 (2003)

9. Asthana, S., King, O.D., Gibbons, F.D., Roth, F.P.: Predicting Protein Complex Membership Using Probabilistic Network Reliability. Genome Res. 14, 1170–1175 (2004)
10. Bader, J.S., Chaudhuri, A., Rothberg, J.M., Chant, J.: Gaining confidence in high-throughput protein interaction networks. Nat. Biotechnol. 22, 78–85 (2004)
11. Sharan, R., Suthram, S., Kelley, R.M., Kuhn, T., McCuine, S., Uetz, P., Sittler, T., Karp, R.M., Ideker, T.: Conserved patterns of protein interaction in multiple species. Proc Natl Acad Sci. USA 102, 1974–1979 (2005)
12. Zhang, L.V., Wong, S.L., King, O.D., Roth, F.P.: Predicting co-complexed protein pairs using genomic and proteomic data integration. BMC Bioinformatics 5, 38 (2004)
13. Marcotte, E.M., Pellegrini, M., Thompson, M.J., Yeates, T.O., Eisenberg, D.: A combined algorithm for genome-wide prediction of protein function. Nature 402, 83–86 (1999)
14. Goldberg, D.S., Roth, F.P.: Assessing experimentally derived interactions in a small world. Proc Natl Acad Sci. USA 100, 4372–4376 (2003)
15. Jansen, R., Yu, H., Greenbaum, D., Kluger, Y., Krogan, N.J., Chung, S., Emili, A., Snyder, M., Greenblatt, J.F., Gerstein, M.: A Bayesian networks approach for predicting protein-protein interactions from genomic data. Science 302, 449–453 (2003)
16. Troyanskaya, O.G., Dolinski, K., Owen, A.B., Altman, R.B., Botstein, D.: A Bayesian framework for combining heterogeneous data sources for gene function prediction (in Saccharomyces cerevisiae). Proc Natl Acad Sci. USA 100, 8348–8353 (2003)
17. von Mering, C., Huynen, M., Jaeggi, D., Schmidt, S., Bork, P., Snel, B.: STRING: a database of predicted functional associations between proteins. Nucleic Acids Res. 31, 258–261 (2003)
18. Walhout, A.J., Sordella, R., Lu, X., Hartley, J.L., Temple, G.F., Brasch, M.A., Thierry-Mieg, N., Vidal, M.: Protein interaction mapping in C elegans using proteins involved in vulval development. Science 287, 116–122 (2000)
19. von Mering, C., Jensen, L.J., Kuhn, M., Chaffron, S., Doerks, T., Kruger, B., Snel, B., Bork, P.: STRING 7–recent developments in the integration and prediction of protein interactions. Nucleic Acids Res. 35, D358–362 (2007)
20. Lee, I., Date, S.V., Adai, A.T., Marcotte, E.M.: A probabilistic functional network of yeast genes. Science 306, 1555–1558 (2004)
21. Letovsky, S., Kasif, S.: Predicting protein function from protein/protein interaction data: a probabilistic approach. Bioinformatics 19, 197–204 (2003)
22. Bader, J.S.: Greedily building protein networks with confidence. Bioinformatics 19, 1869–1874 (2003)
23. Ashburner, M., Ball, C.A., Blake, J.A., Botstein, D., Butler, H., Cherry, J.M., Davis, A.P., Dolinski, K., Dwight, S.S., Eppig, J.T., Harris, M.A., Hill, D.P., Issel-Tarver, L., Kasarskis, A., Lewis, S., Matese, J.C., Richardson, J.E., Ringwald, M., Rubin, G.M., Sherlock, G.: Gene ontology: tool for the unification of biology. The. Gene Ontology Consortium. Nat. Genet. 25, 25–29 (2000)
24. Bader, G.D., Hogue, C.W.: An automated method for finding molecular complexes in large protein interaction networks. BMC Bioinformatics 4, 2 (2003)
25. Spirin, V., Mirny, L.A.: Protein complexes and functional modules in molecular networks. Proc Natl Acad Sci. USA 100, 12123–12128 (2003)
26. Mewes, H.W., Frishman, D., Guldener, U., Mannhaupt, G., Mayer, K., Mokrejs, M., Morgenstern, B., Munsterkotter, M., Rudd, S., Weil, B.: MIPS: a database for genomes and protein sequences. Nucleic Acids Res. 30, 31–34 (2002)
27. Wu, F.Y.: The Potts model. Reviews of Modern Physics 54, 235–268 (1982)
28. Chandler, D.: Introduction to modern statistical mechanics. Oxford University Press, New York (1987)

29. Vazquez, A., Flammini, A., Maritan, A., Vespignani, A.: Global protein function prediction from protein-protein interaction networks. Nat. Biotechnol. 21, 697–700 (2003)
30. Leone, M., Pagnani, A.: Predicting protein functions with message passing algorithms. Bioinformatics 21, 239–247 (2005)
31. Myers, C.L., Robson, D., Wible, A., Hibbs, M.A., Chiriac, C., Theesfeld, C.L., Dolinski, K., Troyanskaya, O.G.: Discovery of biological networks from diverse functional genomic data. Genome Biol. 6, R114 (2005)
32. Scholtens, D., Gentleman, R.: Making sense of high-throughput protein-protein interaction data. Stat Appl Genet Mol Biol. 3 Article39 (2004)
33. Scholtens, D., Vidal, M., Gentleman, R.: Local modeling of global interactome networks. Bioinformatics 21, 3548–3557 (2005)
34. Califano, A., Stolovitzky, G.: DREAM Project. http://magnet.c2b2.columbia.edu/news/DREAMInitiative.pdf

Markov Additive Chains and Applications to Fragment Statistics for Peptide Mass Fingerprinting

Hans-Michael Kaltenbach[1,*], Sebastian Böcker[2], and Sven Rahmann[1,3]

[1] Graduate School in Bioinformatics and Genome Research, Bielefeld University
[2] Lehrstuhl für Bioinformatik, Friedrich-Schiller-University Jena, Ernst-Abbe-Platz 2, D-07743 Jena
[3] Algorithms and Statistics for Systems Biology group, Faculty of Technology, Bielefeld University, D-33594 Bielefeld
michael@cebitec.uni-bielefeld.de

Abstract. Peptide mass fingerprinting is a technique to identify a protein from its fragment masses obtained by mass spectrometry after enzymatic digestion. Recently, much attention has been given to the question of how to evaluate the significance of identifications; results have been developed mostly from a combinatorial perspective. In particular, existing methods generally do not capture the fact that the same amino acid can have different masses because of, e.g., isotopic distributions or variable chemical modifications.

We offer several new contributions to the field: We introduce *probabilistically weighted alphabets*, where each character can have different masses according to a probability distribution, and *random weighted strings* as a fundamental model for random proteins. We develop a general computational framework, *Markov Additive Chains*, for various statistics of cleavage fragments of random proteins, and obtain general formulas for these statistics. Special results are given for so-called standard cleavage schemes (e.g., Trypsin). Computational results are provided, as well as a comparison to proteins from the SwissProt database.

1 Introduction

Mass spectrometry (MS) plays a key role in today's proteomics experiments [1]. The main application of mass spectrometry is the identification of proteins either by de-novo sequencing [2] or by database search, either using peptide-mass fingerprinting [3] or tandem MS.

In peptide-mass fingerprinting, the protein is biochemically digested into fragments using proteases. The set of masses of these fragments, the so-called peptide mass fingerprint (PMF), is measured using MS and compared to a set of reference spectra, usually obtained from in-silico digested database sequences. Different comparison tools have been developed, prominent ones being Mascot [4] based on the MOWSE scoring system [5] and ProFound [6]. One major

---

* Corresponding author.

T. Ideker and V. Bafna (Eds.): Syst. Biol. and Comput. Proteomics Ws, LNBI 4532, pp. 29–41, 2007.
© Springer-Verlag Berlin Heidelberg 2007

problem in developing spectra comparison methods is to estimate the statistical significance of its results. First of all, a statistical model of the digestion and the resulting mass fingerprints is needed.

Presently, there are two major approaches to cope with this problem: (1) Statistical models based on simplifying assumptions, such as the uniform distribution of fragment masses [5], to reduce the problem complexity considerably. (2) Derivation of a model from empirical data, e.g., from large samples of mass spectra (e.g., [7] for tandem MS), where we have to deal with the problem that such data depend on the instruments and their configuration. Another possibility is to use in-silico digestion of whole protein sequence databases [8]. The statistical significance values are then data dependent and are thus hard to compare to other results.

Moreover, we are frequently only interested in the proteome of one species. Deriving an empirical model is often difficult as most species-specific protein databases are too small to get reliable statistical data for fragments and using a very large non-specific database may result in biased estimates.

In [9], a new approach is proposed to compute significance values. Here, the probability that an i.i.d. random sequence contains at least one fragment of certain mass is computed using dynamic programming with uniform character distributions and fixed character masses. However, not all amino acids occur with the same frequency in natural proteins, and the same amino acid may have different masses because of isotopic distributions, or, for example, different variable post-translational modifications that may or may not be present. Therefore, more general models and computational approaches are needed.

This paper provides both a general model, namely *random weighted strings*, which we introduce in Section 2, and a computational framework for several kinds of fragment statistics, namely *Markov Additive Chains* (MACs), which we define and apply in Section 3. The MAC framework is quite general; it comprises, for example, random i.i.d. and Markovian proteins of arbitrary order, and different cleavage rules. Here we focus on i.i.d. strings and *standard cleavage schemes*, to be defined subsequently. We derive the exact distributions of fragment length, the number of fragments, and the joint fragment length-mass distribution. From these results, we derive the probability that a given fragment mass occurs in a string of given length (Section 4). We have implemented our results and compare the i.i.d. model predictions to empirical data obtained from the SwissProt database [10] in Section 5.

Related work. Our results build on three lines of previous research.

First, we extend the concept of *weighted strings* [11], which have been used in the setting of mass spectrometry to generate peptide candidates [12,13], to compute possible decompositions of masses into character masses [14] and to find submasses [15].

Second, the fragment lengths are waiting times for specific, possibly overlapping, patterns in strings. For strings without weights, the statistics of such overlapping patterns [16,17,18] and sets of patterns [19] have been intensively investigated in bioinformatics and statistics [20], and our results on random weighted strings naturally contain some of these as particular cases.

Third, the model of random weighted strings and MACs is a discrete-time variant of an *Markov additive process (MAP)*. MAPs have been studied for continuous-time Markov models and general additive components. Major lines of investigation were existence and limit theorems [21,22], large deviations, and connections to Perron-Frobenius theory [23,24].

Notational conventions. We write $\mathcal{L}(X)$ for the distribution of a random variable (r.v.) X. Distributions are represented as vectors, e.g., we write $x[m] = \mathbb{P}(X = m)$ for some finite range of integers m.

Further, $\mathcal{L}(X) \star \mathcal{L}(Y) = \mathcal{L}(X+Y)$ denotes the convolution of the distributions of two independent r.v.s X and Y. Convolution of two vectors $x[i]$, $y[j]$ is defined as $(x \star y)[k] := \sum_i x[i] \cdot y[k-i]$, where the finite value range of k is derived from the ranges of i and j.

For a string s we denote the substring from index i to index j by $s_{i:j}$, and we write $s^\ell := s_{1:\ell}$.

2 The Random Weighted String Model

We assume Σ to be a finite alphabet. The following definition states the concept of weighted strings as introduced in [11] with an extension from natural to integer masses.

Definition 1 (Weighted alphabet, weighted string, string mass). *Let* $\mu : \Sigma \to \mathbb{Z}$ *be a function assigning each character* $\sigma \in \Sigma$ *its* mass *or* weight $\mu(\sigma) :\equiv \mu_\sigma$. *The pair* (Σ, μ) *is called a* weighted alphabet *with* character mass function μ.

A sequence $(s_i, \mu(s_i))_{i \in \mathbb{N}}$ *on* (Σ, μ) *is called a* weighted string over (Σ, μ). *The marginal sequence* $\mu_i := \mu(s_i)$ *of character masses is the* mass process *of* s.

The mass function is naturally extended to string masses *for finite strings* $s \in \Sigma^*$ *by setting* $\mu(s) := \sum_{i=1}^{|s|} \mu(s_i)$.

In order to capture isotopic distributions and mass modifications of amino acids, we allow multiple masses per amino acid, where each mass is taken with a given probability.

Definition 2 (Probabilistically weighted alphabet (PWA)). *Let* $(\Xi, \mathbb{P})$ *be an appropriately constructed probability space, and let* $\mu : \Sigma \times \Xi \to \mathbb{Z}$ *be a probabilistic character mass function, assigning to each character* $\sigma \in \Sigma$ *a random variable* μ_σ, *so* $\mathbb{P}(\mu_\sigma = m)$ *denotes the probability that the mass of character* σ *takes the value* m. *The pair* (Σ, μ) *is then called a* probabilistically weighted alphabet.

Note that it is sufficient to specify the distribution $\mathcal{L}(\mu_\sigma)$ for each $\sigma \in \Sigma$, so we do not need to explicitly specify the probability space Ξ. Also, for $\mathcal{L}(\mu_\sigma)$ Dirac, the PWA is identical with a weighted alphabet.

Since we consider strings of arbitrary length in what follows, we develop our models for a semi-infinite string $s \in \Sigma^{\mathbb{N}}$ and then use projections to finite length-ℓ prefixes s^{ℓ} as needed. We assume that the masses of characters at different positions are conditionally independent, given the characters.

As for deterministic weighted strings, we define the string mass as the sum of its character masses, but it is now a random variable whose distribution can be computed as the convolution of the character distributions. For a deterministic weighted alphabet, this coincides again with Definition 1.

Lemma 1 (String mass distribution). *For finite $I := \{i_1, i_2, \ldots, i_n\} \subset \mathbb{N}$, let $s_I := s_{i_1} s_{i_2} \ldots s_{i_n}$. The distribution of the string string mass of s_I is given by $\mathcal{L}(\mu(s_I)) = \mathcal{L}(\mu_{i_1}) \star \cdots \star \mathcal{L}(\mu_{i_n})$.*

Up to now, we do not assume a random model $\mathcal{L}(S)$ for a string S over an alphabet; we now show how to combine standard models with weighted strings to derive random weighted strings.

Definition 3 (Random weighted string). *A random weighted string is a stochastic process $(S, \mu) = ((S_1, \mu_1), (S_2, \mu_2), \ldots)$ with index set $\mathbb{N}$, values in $\Sigma \times \mathbb{Z}$ and finite dimensional distributions $\mathcal{L}((S, \mu)_I) = \mathcal{L}(S_I) \otimes \mathcal{L}(\mu_I)$, where S is a random string and μ is the mass process associated to S.*

Henceforth, we discuss the i.i.d. string model, but all above definitions capture arbitrary string models, the most prominent one being Markov sequences.

Given amino acid frequencies, which can be estimated from sequence databases such as SwissProt [10], and the isotopic distribution of amino acids, which can be computed from the isotopic distributions of their component atoms, we can model proteins as random weighted strings.

3   Fragmentation of Random Weighted Strings

Proteases usually cleave right after the occurrence of a specific character. The re-action can be suppressed, however, if this *cleavage character* is directly followed by a so-called *prohibition character*. The digestion process induces a *fragmenta-tion* of the protein sequence; this is formalized below.

Our derivation first focuses on semi-infinite strings $S \in \Sigma^{\mathbb{N}}$ to avoid complica-tions with boundary effects; the necessary adjustments are made subsequently. This is reflected in our notation as follows. Whenever we adjust a quantity, e.g., L_i for the length of the i-th fragment, to strings of finite length ℓ, we denote the adjusted random variable by the superscripted string length, e.g., L_i^{ℓ}. We write $\bar{\Gamma}$ for the complement of a set $\Gamma \subset \Sigma$ in Σ, i.e., $\bar{\Gamma} := \Sigma \setminus \Gamma$.

3.1   Cleavage Schemes and Markov Additive Chains

We introduce the general *Markov Additive Chain* (MAC) framework to carry out computations on statistics of proteins and their fragmentations. We start by defining a cleavage scheme, which describes the activity of many peptide-cleaving

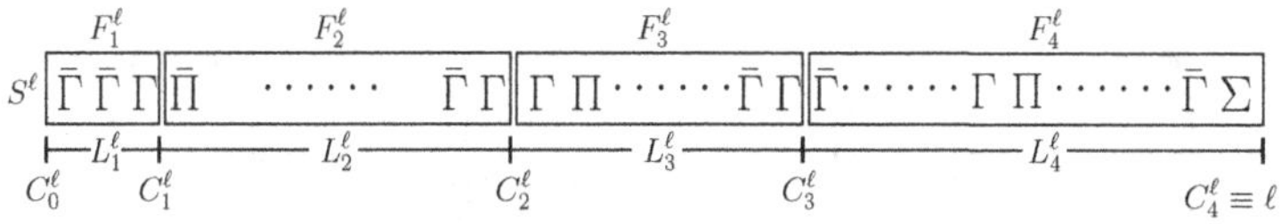

Fig. 1. Fragments F_i, cleavage points T_i and fragment length L_i of a string S

enzymes, and naturally leads to a MAC model subsequently. Applying a cleavage scheme on a string results in a fragmentation of this string in consecutive, non-overlapping substrings, the fragments.

Definition 4 (Cleavage scheme (Γ, Π), Quantities γ, π). *A cleavage scheme is a pair (Γ, Π) of a set of cleavage characters $\Gamma \subset \Sigma$, and a set of prohibition characters $\Pi \subset \Sigma$. If the additional constraint $\Gamma \cap \Pi = \emptyset$ (i.e., $\Gamma \subset \bar{\Pi}$) holds, we speak of a standard cleavage scheme. Strings $C = C_1 C_2 \in \Gamma \bar{\Pi}$ are called cleavage patterns. We set $\gamma := \mathbb{P}(S_i \in \Gamma)$, $\pi := \mathbb{P}(S_i \in \Pi)$.*

Standard schemes are simple enough to allow closed formulas for certain probability distributions (see below), and also powerful enough to capture many real enzymes. For example, the frequently used protease Trypsin cleaves after lysine (K) or arginine (R), unless followed by proline (P). Thus $\Gamma = \{K, R\}$ and $\Pi = \{P\}$.

Definition 5 (Cleavage points). *Define the sequence $(T_i(S)) \equiv (T_i)_{i \in \mathbb{N}}$ of cleavage points of S by $T_0 := 0$ and for $i \geq 1$,*

$$T_i := \min\{k > T_{i-1}(S) : S_k \in \Gamma, S_{k+1} \in \bar{\Pi}\},$$

where we set $T_i := \infty$ if the minimum is taken over the empty set.

For finite length prefixes S^ℓ, we define $T_i^\ell := \min\{T_i, \ell\}$, so that eventually all cleavage points lie directly behind the end of the prefix. We also call $N^\ell := \min\{k : T_k = \ell\}$ the fragmentation size of S^ℓ, as it gives the number of its fragments.

Definition 6 (Fragments, fragmentation). *For each $i \geq 1$, the substring $F_i := S_{T_{i-1}+1:T_i}$ is called the i-th fragment of S. If $T_{i-1} = T_i$, the i-th fragment and the following fragments are empty. We denote the length of fragment F_i by $L_i := T_i - T_{i-1}$. The family $\mathcal{F} := (F_i)_{i \geq 1}$ is called the fragmentation of S. For finite strings S^ℓ, we define F_i^ℓ, L_i^ℓ, and $\mathcal{F}^\ell$ analogously in terms of T_i^ℓ.*

Markov Additive Chains (MACs). MACs generate protein fragment sequences together with their mass process.

Definition 7 (MAC). *A Markov Additive Chain (MAC) is a 6-tuple $(N, \Sigma, P, p^0, Q, F)$ consisting of a finite set of states N, a finite output alphabet Σ, a (sub-)stochastic state transition matrix $P = (P_{ij})_{i,j \in N}$, a start distribution p^0, a family $Q = (Q_i)_{i \in N}$ of output distributions of weighted characters $Q_i = (q_{i,(\sigma,m)})_{(\sigma,m) \in \Sigma \times \mathbb{Z}}$, and a set of final states $F \subset N$.*

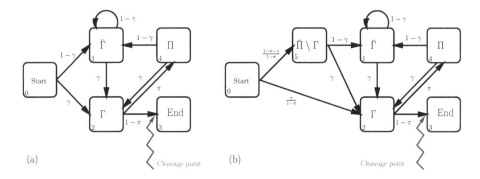

Fig. 2. (a) MAC generating an initial fragment of a random i.i.d. string under a standard cleavage scheme; (b) dito, for an inner fragment. States are labeled with a number i and a character set $A \subset \Sigma$. A MAC generates a fragment as follows: Starting in the start state numbered 0, it picks a transition according to the probability distribution of the state's outgoing edges and then emits a weighted character from the new state's character set A according to its conditional joint character-mass distribution.

The semantics are as follows: The start state i is picked according to distribution p^0. A transition to a new state, being in state i, is made according to the probability distribution in row i of P. For states $i \in F$, these distributions are defective (i.e., they do not sum to one); the MAC halts in these states. The sequence of states taken through the MAC thus forms a *Markov chain* with transition matrix P. When state i is entered after a transition, a random weighted character (C, μ) is output according to the joint character-mass distribution Q_i. We assume that the output characters of the final states to not belong to the fragment sequence (i.e., cleavage occurs before entering F), but we do allow masses in these states to model certain groups at the end of a fragment.

Constructing MACs from a standard cleavage scheme. Two particular examples of MACs, constructed from a standard cleavage scheme (Γ, Π), are shown in Figure 2. The MAC in Figure 2a generates an initial fragment F_1, while the MAC in Figure 2b generates a subsequent fragment F_k, $k \geq 2$.

The state set N, the transition matrix P, and the final state set $F = \{3\}$ are obvious from the figure. The start state is alway state 0, so $p^0 = (1, 0, \ldots, 0)$. The alphabet Σ is the amino acid alphabet, plus possibly, initial and terminal symbols that are output in states 0 and 3, respectively, that do not contribute visibly to the protein sequence, but may contribute to its mass by specifying appropriate mass distributions in these states.

The distributions Q_i are constructed for $i \notin \{0, 3\}$ from the state-labeling character subsets A_i in Figure 2: We set $q_{i,(\sigma,m)} := \mathbb{P}(C = \sigma, \mu = m \mid C \in A_i)$. Note that the transition probabilities *into* state i always equal $\mathbb{P}(C \in A_i)$.

The framework of MACs is much more general: We can construct MAC models for more complicated cleavage rules and string models. We do not pursue this here further, but consider probability computations with MACs instead.

3.2   Fragment Length Distribution

Since the first fragment has a different prefix than the following ones, but all following ones are i.i.d., we define

$$u_1[l] := \mathbb{P}(L_1 = l), \quad \text{and} \quad u_+[l] := \mathbb{P}(L_+ = l),$$

where L_+ stands for any L_k for $k \geq 2$. We compute the length distributions by following paths through the MACs in Figure 2. In fact, the following result does not make use of the weight distributions and is well known from Markov chain theory.

Theorem 1 (Fragment Length distribution). *We have*

$$u_\circ[l] = \sum_{i \in F} [p^0 \cdot P^{l+1}]_i,$$

where $\circ \in \{1, +\}$ *depending on whether an initial or subsequent fragment is considered.*

Proof. Let p^l denote the state distribution after l steps, then $p^l = p^0 \cdot P^l$ by the Chapman-Kolmogorov equation. For fragment length exactly l, we need to be in a final state F after $l + 1$ steps.

For the models in Figure 2, we obtain closed formulas for $u_\circ[l]$, and, using generating functions, also for the moments of the distribution. The details are omitted here and can be found in a technical report [25]. Numerical results follow in Section 5.

Finite strings. For finite string length ℓ, define

$$u_1^\ell[l] := \mathbb{P}(L_1^\ell = l), \text{ and } u_+^\ell[l] := \mathbb{P}(L_i^{\ell+k} = l \mid L_{i-1} = k) \text{ for any } i \geq 2 \text{ and any } k \in \mathbb{N}.$$

The second definition is in fact independent of i and k, and simply defines the conditional distribution of L_i given that there are ℓ characters left in the string. See [25] for the necessary adjustments to compute $u_\circ^\ell$.

Approximation. The length distributions are the waiting times for the cleavage pattern with probability $\gamma(1 - \pi)$; they are approximated by a geometric distribution with parameter $p = \gamma(1 - \pi)$. Computing moments, more precise parameters are $p = 1/\mathbb{E}(L_1)$, $p = 1/\mathbb{E}(L_+)$, respectively.

3.3   Number of Fragments

From the distributions for the fragment length we derive the exact distribution for the number of fragments N^ℓ.

Lemma 2 (Relationship of N^ℓ and T_k). N^ℓ *is related to the cleavage points* T_k *by* $\mathbb{P}(N^\ell \leq k) = \mathbb{P}(T_k \geq \ell)$ *and* $\mathbb{P}(N^\ell = k) = \mathbb{P}(N^\ell \leq k) - \mathbb{P}(N^\ell \leq k - 1) = \mathbb{P}(T_k \geq \ell) - \mathbb{P}(T_{k-1} \geq \ell).$

Lemma 3 (Exact distribution of cleavage points).

$$\mathcal{L}(T_k) = \mathcal{L}(L_1) \star \mathcal{L}(L_2)^{\star(k-1)} = \mathcal{L}(T_{k-1}) \star \mathcal{L}(L_k).$$

Proof. Remember that $T_k = \sum_{i=1}^k L_i$ and that the L_i are independent.

Approximation. The distribution of cleavage point T_k is a sum of k nearly geometric distributions each starting at one. It corresponds to a negative binomial distributions with size parameter k and probability parameter $p = k/(\mathbb{E}(L_1) + (k-1)\mathbb{E}(L_+) + k)$, where we have to add k in the denominator to shift to geometric distributions starting at zero.

3.4 Length-Mass Distributions

We now examine the joint distribution of length and mass of fragments of semi-infinite random weighted i.i.d. strings (S, μ). We define

$$f_1[l, m] := \mathbb{P}(L_1 = l, \mu_{F_1} = m), \text{ and } f_+[l, m] := \mathbb{P}(L_i = l, \mu_{F_i} = m) \text{ for any } i \geq 2.$$

The definition of f_+ is independent of i because all fragments of S except the first are i.i.d.

These distributions can be computed efficiently using the MAC framework in Figure 2, and in fact we have designed MACs to solve this particular problem.

Lemma 4 (Mass added in state i). *Let $g_i[m]$ be the probability that a character generated in state i has mass m. Then we have $g_i[m] = \sum_{\sigma \in \Sigma} q_{i,(\sigma,m)}$.*

For the special case of standard cleavage schemes and i.i.d. strings (see the MAC construction in Section 3.1), let A_i be the character set associated to state $i \notin \{0, 3\}$ (see Figure 2), and let C denote a random character of the i.i.d. model. Then $g_i[m] = \frac{1}{\mathbb{P}(C \in A_i)} \cdot \sum_{c \in A_i} \mathbb{P}(C = c) \cdot \mathbb{P}(\mu_c = m)$. In other words, the mass distribution in state i is a mixture of the $|A_i|$ mass distributions $\mathcal{L}(\mu_c)$ with mixture coefficients $\mathbb{P}(C = c)/\mathbb{P}(C \in A_i)$ for $c \in A_i$. For the MACs in Figure 2, we have a Dirac distribution at mass zero for the start state 0 and the final state 3, because the start state does not contribute any mass and cleavage occurs before state 3. However, if we want to model additional chemical groups that are attached before and after each fragment, we can use these states to model arbitrary mass distributions for these groups.

Theorem 2 (Computation of f). *Let $h_i^l[m]$ be the probability that, after l steps, we are in state i and the cumulative mass of the fragment generated so far is m. Let us define column vectors $g_i := (g_i[m])_m$ and $h_i^l := (h_i^l[m])_m$ matrices $G := (g_1 | \ldots | g_n)$ and $H^l := (h_1^l | \ldots | h_n^l)$.*
Then initially, $h_i^0 = p_i^0 \cdot g_i$ for $i \in N$, and after step $l \geq 1$,

$$H^l = (H^{l-1} \cdot P) \star G, \tag{1}$$

where we define convolution in a column-by-column manner:

$$X \star G \equiv (x_1 | \ldots | x_n) \star (g_1 | \ldots | g_n) := (x_1 \star g_1 | \ldots | x_n \star g_n).$$

Finally, $f_\circ[l, m] = \sum_{i \in F} h_i^{l+1}[m]$.

Table 1. Derivation of $f_\circ[l, m]$ for i.i.d. strings and standard cleavage schemes

$f_1[l, m]$	$f_+[l, m]$
$H^0 = (g_0, 0, \ldots, 0)$	$H^0 = (g_0, 0, \ldots, 0)$
$h_0^l = 0$	$h_0^l = 0$
$h_1^l = (1 - \gamma) \cdot (h_0^{l-1} + h_1^{l-1} + h_4^{l-1}) \star g_1$	$h_1^l = (1 - \gamma) \cdot (h_1^{l-1} + h_4^{l-1} + h_5^{l-1}) \star g_1$
$h_2^l = \gamma \cdot (h_0^{l-1} + h_1^{l-1} + h_4^{l-1}) \star g_2$	$h_2^l = \gamma \cdot [1/(1 - \pi) \cdot h_0^{l-1} + h_1^{l-1} + h_4^{l-1} + h_5^{l-1}] \star g_2$
$h_3^l = (1 - \pi) \cdot h_2^{l-1} \star g_3$	$h_3^l = (1 - \pi) \cdot h_2^{l-1} \star g_3$
$h_4^l = \pi \cdot h_2^{l-1} \star g_4$	$h_4^l = \pi \cdot h_2^{l-1} \star g_4$
	$h_5^l = (1 - \pi - \gamma)/(1 - \pi) \cdot h_0^{l-1} \star g_5$
$f_1[l, m] = h_3^{l+1}[m]$	$f_1[l, m] = h_3^{l+1}[m]$

Proof. The initial conditions are obvious (cf. also the above remark). To compute $h_i^l[m]$ for $l \geq 1$, consider the possible states k in step $l - 1$, their transition probabilities P_{ki} to state i and the possible masses m' accumulated in step $l-1$. We obtain

$$h_i^l[m] = \sum_{m'} \sum_{k} \mathbb{P}(\text{in state } k \text{ after } l - 1 \text{ steps}, \text{transition to } i, \text{mass added is } m')$$

$$= \sum_{m'} \left(\sum_{k} P_{ki} \cdot h_k^{l-1}[m - m'] \right) \cdot g_i[m'] = [(H^{l-1} \cdot P) \star G]_{m,i};$$

thus $H^l = (H^{l-1} \cdot P) \star G$ as claimed.

Corollary 1 (Computation of f for i.i.d. strings and standard cleavage schemes). *Applying 2 to the MACs in Figure 2, we obtain the derivation for $f_\circ[l, m]$ for i.i.d. strings and standard cleavage schemes shown in Table 1.*

Finite strings. As for the length distributions, we denote length-mass distribution for finite string length as follows:

$$f_1^\ell[l, m] := \mathbb{P}(L_1^\ell = l, \mu_{F_1^\ell} = m),$$

$$f_+^\ell[l, m] := \mathbb{P}(L_i^{\ell+k} = l, \mu_{F_i^{\ell+k}} = m \mid T_{i-1} = k) \text{ for any } i \geq 2 \text{ and any } k \in \mathbb{N}.$$

The second definition is in fact independent of i and k, and defines the conditional join distribution of (L_i, μ_{F_i}) given that there are ℓ characters left in the string.

The necessary adjustments can again be found in [25].

Mass avoidance probabilities. To conclude this derivation, we state the obvious fact that we can also compute the probability that a fragment has length l and *not* mass m as $\bar{f}_\circ[l, m] = u_\circ[l] - f_\circ[l, m]$, and $\bar{f}_\circ^\ell[l, m] = u_\circ^\ell[l] - f_\circ^\ell[l, m]$.

4   Occurrence of Masses in Random Weighted Strings

To estimate the significance of a protein identification, utilization of the probabilities that a string S^ℓ has a fragment of mass m (and thus gives a signal at that mass in the mass spectrum) is often required.

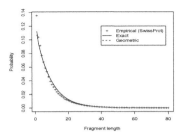
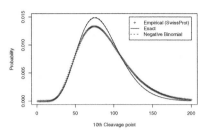

Fig. 3. Empirical SwissProt and exact densities for L_+ (left, additional geometric) and T_{10} (right with negative binomial)

Let $N^\ell(m) := \left|\{F \in \mathcal{F}_S^\ell | \mu_F = m\}\right|$ be the number of fragments of mass m in S^ℓ and define the *mass occurrence probability* and its complement as

$$p^\ell[m] := \mathbb{P}(N^\ell(m) \geq 1) \quad \text{and} \quad \bar{p}^\ell[m] := 1 - p^\ell[m].$$

Similarly, define $p_+^L[m]$ and its complement $\bar{p}_+^L[m]$ as the corresponding probabilities for a string of length L that consists homogeneously of subsequent fragments (i.e., without considering the first fragment).

Lemma 5 (Mass occurrence). *The occurrence probabilities are recursively given by*

$$\bar{p}^\ell[m] = \sum_{l=1}^{\ell} \bar{p}_+^{\ell-l}[m]\bar{f}_1^\ell[l,m] \quad \text{and} \quad \bar{p}_+^L[m] = \sum_{l=1}^{L} \bar{p}_+^{L-l}[m]\bar{f}_+^L[l,m]$$

with the obvious initial condition $\bar{p}^0[m] = 1$ and $\bar{p}_+^0[m] = 1$ for $m \neq 0$.

Proof. The main observation for the proof is that although the fragment masses are not independent, as we deal with finite string length, the mass of the first fragment becomes independent of the remaining fragments' masses if it is known that $L_1 = l$; thus we condition on L_1. The argument can be repeated for the resulting suffix of S^ℓ.

5   Results and Discussion

We compared the results of our model using in-silico digestion of the SwissProt database release 48.0 with Trypsin. The parameters $\gamma = 0.1125$ and $\pi = 0.0483$ were estimated from the amino acid frequencies.

Figure 3 (left) shows a good agreement of the theoretical length distribution $u_+[l]$ with the derived empirical distribution and the approximating Geom $(1/\mathbb{E}(L_+))$. We may also compute moments as $\mathbb{E}(L_1) = 9.391 \pm 8.882$, $\mathbb{E}(L_+) = 9.340 \pm 8.879$ which can be compared to the corresponding $G \sim \text{Geom}(\gamma(1-\pi))$ moments, $\mathbb{E}(G) = 9.340 \pm 8.826$.

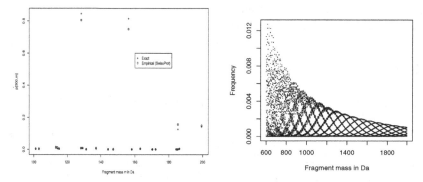

Fig. 4. $p^\ell[m]$ for $\ell = 300$, precision $\Delta = 0.1$, tryptic digestion. Left: Only $p^\ell[m] > 0.001$ shown for mass range 100–300 Da. Right: All probabilities for mass range 600–2000 Da; note that this is *one* (heavily fluctuating) function.

In Figure 3 (right), we compared the distribution of the 10th cleavage point with the empirical SwissProt counterpart and the approximating negative binomial distribution. The approximation is good, whereas the not exact match between short empirical and theoretical fragment lengths also cause the corresponding cleavage point distributions to differ.

In Figure 4 (left), the theoretical occurrence probabilities are compared to their empirical counterparts, again showing a reasonable agreement. Both the theoretical and the empirical model give high probability to the occurrence of masses for certain two-character fragments (> 0.2) as well as for K and R themselves (≈ 0.8) in the range of $100 \ldots 300$ Da. For better presentation, only probabilities greater 0.001 are shown. For higher mass range, the theoretical occurrence probabilities clearly show a combinatorial behavior caused by the length-mass distributions. These probabilities are highly dependent on the possible amino acid combinations to achieve a certain mass.

In conclusion, we presented Markov Additive Chains, a general computational framework for length and mass statistics of random weighted strings and their fragments. Our main result for MACs is Theorem 2 and Equation (1) in particular. We showed in detail how the distributions of these statistics can be derived from the model for the case of i.i.d. strings and standard cleavage schemes. The model is readily extendible to Markov sequences and general cleavage schemes. Mass modifications of the sequence termini can easily be modeled, and it is also possible to introduce probabilistically missed cleavage sites. We also gave and justified several approximations to the distributions not involving masses. Comparisons to empirical data derived from the SwissProt database showed a reasonable agreement but also some discrepancies which are likely to be caused by an oversimplification due to the i.i.d. string model. Further analysis with Markov sequences in currently under way.

Acknowledgments. The authors would like to thank Henner Sudek for help in the implementation of the algorithms and for running the experiments together with Matthias Steinrücken. Steffen Grossmann and Jens Stoye have provided valuable input and discussion to clarify the subject.

References

1. Aebersold, R., Mann, M.: Mass spectrometry-based proteomics. Nature 422, 198–207 (2003)
2. Frank, A., Pevzner, P.: PepNovo: de novo peptide sequencing via probabilistic network modeling. Anal Chem. 15, 964–973 (2005)
3. Henzel, W.J., Watanabe, C., Stults, J.T.: Protein idendification: The origins of peptide mass fingerprints. J.Am.Soc.Mass Spectrometry 14, 931–942 (2003)
4. Perkins, D., Pappin, D., Creasy, D., Cottrell, J.: Probability-based protein identification by searching sequence databases using mass spectrometry data. Electrophoresis 20, 3551–3567 (1997)
5. Pappin, D., Hojrup, P., Bleasby, A.: Rapid identification of proteins by peptide-mass fingerprints. Current Biology 3, 327–332 (1993)
6. Zhang, W., Chait, B.T.: ProFound: an expert system for protein identification using mass spectrometric peptide mapping information. Anal. Chem. 72, 2482–2489 (2000)
7. Colinge, J., Masselot, A., Magnin, J.: A systematic statistical analysis of ion trap tandem mass spectra in view of peptide scoring. In: Benson, G., Page, R.D.M. (eds.) WABI 2003. LNCS (LNBI), vol. 2812, pp. 25–38. Springer, Heidelberg (2003)
8. Wang, I.J., Diehl, C.P., Pineda, F.J.: A statistical model of proteolytic digestion. In: Proceedings of IEEE CSB 2003, Stanford, California pp. 506–508 (2003)
9. Böcker, S., Kaltenbach, H.M.: Mass spectra alignments and their significance. In: Apostolico, A., Crochemore, M., Park, K. (eds.) CPM 2005. LNCS, vol. 3537, pp. 429–441. Springer, Heidelberg (2005)
10. Bairoch, A., Boeckmann, B.: The SWISS-PROT protein sequence data bank. Nucleic Acids Res. 20, 2019–2022 (1992)
11. Cieliebak, M., Erlebach, T., Lipták, Zs., Stoye, J., Welzl, E.: Algorithmic complexity of protein identification: Combinatorics of weighted strings. Discrete Applied Mathematics 137, 27–46 (2004)
12. Edwards, N., Lippert, R.: Generating peptide candidates from amino-acid sequence databases for protein identification via mass spectrometry. In: Guigó, R., Gusfield, D. (eds.) WABI 2002. LNCS, vol. 2452, pp. 68–81. Springer, Heidelberg (2002)
13. Wan, Y., Chen, T.: A hidden markov model based scoring function for mass spectrometry database search. In: Miyano, S., Mesirov, J., Kasif, S., Istrail, S., Pevzner, P., Waterman, M. (eds.) Research in Computational Molecular Biology. LNCS (LNBI), vol. 3500, pp. 342–356. Springer, Heidelberg (2005)
14. Böcker, S., Lipták, Zs.: Efficient mass decomposition. In: Proc. of ACM Symposium on Applied Computing (ACM SAC 2005), Santa Fe, USA, pp. 151–157 (2005)
15. Bansal, N., Cieliebak, M., Lipták, Zs.: Efficient algorithms for finding submasses in weighted strings. In: Sahinalp, S.C., Muthukrishnan, S.M., Dogrusoz, U. (eds.) CPM 2004. LNCS, vol. 3109, pp. 194–204. Springer, Heidelberg (2004)
16. Régnier, M.: A unified approach to word occurrence probabilities. Discrete Applied Mathematics 104, 259–280 (2000)
17. Waterman, M.S.: Introduction to Computational Biology, 1st edn. CRC Press, Boca Raton (1996)

18. Reinert, G., Schbath, S., Waterman, M.S.: Probabilistic and statistical properties of words: An overview. Journal of Computational Biology 7, 1–46 (2000)
19. Robin, S., Daudin, J.J.: Exact distribution of word occurrences in a random sequence of letters. Journal of Applied Probability 36, 179–193 (1999)
20. Wyner, A.J.: More on recurrence and waiting times. The Annals of Applied Probability 9, 780–796 (1999)
21. Cinlar, E.: Markov Additive Processes I. Z. Wahrscheinl. verw. Geb. 24, 85–93 (1972)
22. Cinlar, E.: Markov Additive Processes II. Z. Wahrscheinl. verw. Geb. 24, 95–121 (1972)
23. Ney, P., Nummelin, E.: Markov Additive Processes I. Eigenvalue properties and limit theorems. Ann. Probab. 15, 561–592 (1987)
24. Ney, P., Nummelin, E.: Markov Additive Processes II. Large deviations. Ann. Probab. 15, 593–609 (1987)
25. Kaltenbach, H.M., Sudek, H., Böcker, S., Rahmann, S.: Statistics of cleavage fragments in random weighted strings. Technical Report 2005-06, Technische Fakultät der Universität Bielefeld, Abteilung Informationstechnik (2005)

A Context-Specific Network of Protein-DNA and Protein-Protein Interactions Reveals New Regulatory Motifs in Human B Cells

Celine Lefebvre[1], Wei Keat Lim[1], Katia Basso[2], Riccardo Dalla Favera[2], and Andrea Califano[1,*]

[1] Center for Computational Biology and Bioinformatics, Columbia University
1130 St Nicholas Avenue, 9th Floor, New York NY 10032
califano@c2b2.columbia.edu
[2] Institute of Cancer Genetics, Columbia University
1130 St Nicholas Avenue, New York NY 10032

Abstract. Recent genome wide studies in yeast have started to unravel the complex, combinatorial nature of transcriptional regulation in eukaryotic cells, including the concerted regulation of proteins involved in complex formation. In this work, we use a Bayesian evidence integration framework to assemble an integrated network, including both protein-DNA and protein-protein interactions, in a specific cellular context(human B cells). We then use it to study common interaction motifs between protein complexes and regulatory programs, using an enrichment analysis approach. Specifically, we compare the frequency of mixed interaction motifs in the real network against random networks of equivalent connectivity. These motifs include sets of target genes regulated by multiple interacting transcription factors, and gene sets encoding same complex proteins regulated by different transcription factors.

Keywords: combinatorial regulation / evidence integration / human B cells / naïve Bayes / network motifs.

1 Introduction

Dissecting transcriptional regulation pathways in mammalian cells is an important step towards the elucidation of normal and disease-related cellular processes. Due to its simpler organization, yeast has so far provided an excellent model organism for the study of eukaryotic cellular networks, offering an initial basis to understand their dynamic complexity. Recently, motifs analyses in yeast networks combining Protein-DNA (P-D) and Protein-Protein (P-P) interactions revealed a trend towards co-regulation and complex formation in lower eukaryotes [1,2], showing that the integration of different interaction types helps elucidate the interface between transcriptional regulation and protein complex formation.

---

* Corresponding author.

T. Ideker and V. Bafna (Eds.): Syst. Biol. and Comput. Proteomics Ws, LNBI 4532, pp. 42–56, 2007.

Similar models have not yet become available for higher eukaryotes, including *Homo sapiens*, where transcriptional regulation, complex-formation, and transient protein-protein interactions networks have been studied in isolation and without cell context specificity, for instance by yeast two-hybrids [3,4]. Here, we propose a Bayesian evidence integration framework for network inference, which integrates a variety of generic and context specific experimental clues about P-P and P-D interactions - such as a large collection of B cell expression profiles - with inferences from different reverse engineering algorithms, such as GeneWays [5] and ARACNE [6]. This type of Bayesian learning was previously successful in inferring P-P interactions in yeast [7] and in human [8]. The resulting network is then used as a model to study the interface between regulatory programs and protein complexes.

We first analyzed the enrichment of simple three gene motifs involving both P-P and P-D interactions in our network and then combined them into larger composite motifs to identify combinatorial regulation mechanisms.

2 Material and Methods

2.1 Naïve Bayesian Evidence Integration

The Bayesian evidence integration model applies the Bayes theorem to compute the posterior odds that a specific interaction exists (O_{post}) as the product of the prior odds (O_{prior}) and of a likelihood ratio (LR) [7]:

$$O_{post} = LR \cdot O_{prior} \tag{1}$$

The prior odds are defined as the average odds that two random gene products are involved in an interaction and can be calculated as:

$$O_{prior} = \frac{P(I)}{P(\bar{I})} = \frac{P(I)}{1 - P(I)} \tag{2}$$

where $P(I)$ is the probability that two random gene products are involved in an interaction and $P(\bar{I})$ is the probability that they are not. The posterior odds of a specific interaction are defined as the ratio of the probabilities that two specific gene products, g_x and g_y, are respectively involved or not involved in an interaction, conditional to the presence of N different clues, $c_1...c_N$:

$$O_{post} = \frac{P(I_{xy}|c_1 \cdots c_N)}{P(\bar{I}_{xy}|c_1 \cdots c_N)} \tag{3}$$

Such clues could include, for instance, the correlation of the two genes' expression profiles, the results of specific experimental assays, the functional categorization of the gene pair, etc. Similarly, the LR is defined as:

$$LR(c_1 \cdots c_N) = \frac{P(c_1 \cdots c_N|I_{xy})}{P(c_1 \cdots c_N|\bar{I}_{xy})} \tag{4}$$

In the Naïve Bayes Classifier (NBC) model, the clues are assumed to be statistically independent. Then, the LR can be computed as the product of individual LR from the respective datasets:

$$LR(c_1 \cdots c_N) = \prod_{i=1}^{i=N} LR(c_i) = \prod_{i=1}^{i=N} \frac{P(c_i|I_{xy})}{P(c_i|\bar{I}_{xy})} \qquad (5)$$

A useful property of the NBC model is that performance does not significantly deteriorate if weak dependencies among the clues exist. Under this assumption, the posterior odds of a specific interaction can be calculated as:

$$O_{post} = \prod_{i=1}^{i=N} \frac{P(c_i|I_{xy})}{P(c_i|\bar{I}_{xy})} \cdot O_{prior} \qquad (6)$$

O_{prior} can be estimated from prior knowledge on the number of expected P-P interactions or P-D interactions in a cellular context, while the LRs are estimated by counting how many times a specific clue is observed in a positive and negative *gold standard* set. A positive gold standard set should include only gene product pairs that are known to interact, while a negative gold standard set should include only gene product pairs that are known not to interact. O_{post}, computed as the product of these two values, is related to the probability of an interaction to be true as $P_{post} = O_{post}/(O_{post} + 1)$, then achieving a posterior probability of at least 50% is equivalent to achieve $O_{post} \geq 1$ or $LR \geq 1/O_{prior}$.

2.2   Gold-Standard Sets for P-P Interactions

To generate a positive gold standard set (GSP) for P-P interactions, we extracted 25,642 human P-P interactions from HPRD [9], 7,862 from IntAct [10], 4,812 from BIND [11], and 868 from DIP [12], originating from low-throughput, high quality experiments. This resulted in a GSP set of 34,842 unique P-P interactions involving 7,323 genes (28,542 interactions for 6,953 genes after homodimers removal). Based on an estimate for the total number of P-P interactions of 300,000 in a human cell, among 22,000 proteins [3], the prior odds for an interaction is approximately 1 in 800 $(300,000/(22,000^2/2 - 300,000))$. This implies that any protein pair with a $LR \geq 800$ has at least a 50% probability of being involved in an interaction. Generating a negative gold standard set (GSN) is somewhat more complicated because negative interaction examples are not easily identified from the literature. Thus, based on a previous analysis [13], we classified the Gene Ontology (GO) terms into four subcellular compartments (cell periphery and exocytic pathway, cytoplasm, mitochondria and nucleus), and mapped human genes into those compartments. Then, for each compartment pair, we computed the enrichment of protein pairs known to interact (from the GSP) using a Fisher exact test (FET). This revealed compartment pairs that are more likely to host proteins involved in a P-P interaction. Obviously, all pairs where the two compartments were identical (e.g., nucleus-nucleus) showed enrichment.

However, interestingly, the pair cytoplasm-mitochondrion also showed enrichment. The GSN was then defined by using all proteins from non-enriched cell compartment pairs. Note that we also excluded nucleus-mitochondrion protein pairs in the GSN, as the FET was borderline. This resulted in a GSN with 18,359,948 candidate non interacting gene pairs. As could be expected, the GSN had a small overlap with the GSP (1,890 pairs) reflecting the heuristic nature of the approach used to identify negative interactions. However, the overlap is much smaller than expected by chance thus validating that the method provides a relatively good first order approximation of non interacting protein pairs GSN. For our subsequent analysis, we removed from the GSN all the pairs that were also present in the GSP.

2.3 Gold-Standard Sets for P-D Interactions

Defining a realistic GSP for P-D interactions is much more difficult, as the amount of biochemically validated data is orders of magnitude smaller. We thus decided to focus on a very well-studied transcription factor (TF), MYC for which extensive binding data was collected. The GSP was thus defined as a set of 1,041 B cell specific MYC target genes collected from the MYC database [14] and the GSN as its genomic complement. This allowed us to estimate the prior odds for a MYC P-D interaction to be 1 in 21. This causes two problems: First this is likely an underestimate of the total number of MYC targets in a B cell, thus resulting in a corresponding underestimate of the LR for MYC interactions. Second, this LR should not be used for other TFs that may have a smaller or larger number of targets. However, since this data is not available, we used this value as a first order approximation from all TFs. The LR can be iteratively corrected on a TF by TF basis either by estimating the number of actual targets (e.g. by using general properties of the network connectivity [2,15]) or as additional biochemical evidence emerges, such as from ChIP-Chip data.

2.4 Gene Expression Profiles

A collection of 254 gene expression profiles was used, representing 27 distinct cellular phenotypes derived from populations of normal and neoplastic human B lymphocytes [16]. Gene expression profiles were collected using the Affymetrix HG-U95A GeneChip®System (approximately 12,600 probe sets). Expression measurements were normalized using MAS5.0, and probe sets with absolute expression mean < 50 and coefficient of variation < 0.3, were considered non-informative and were excluded a-priori from the analysis, leaving 7,476 probe sets [15]. We computed the mutual information (MI) between the 7,896 probes (6,083 genes) passing this threshold. Mutual Information [6] is an optimal measure of statistical dependence in a non linear setting. After applying a threshold (MI ≥ 0.069), corresponding to a p-value of 10^{-7}, we identified 4,711,682 statistically significant MIs between the 6,083 genes. The highest MI among all the probe-set pairs corresponding to a gene pair was used when multiple probes were present in the set.

2.5   Information Content

As previously described [17], we computed the information content of a Gene Ontology (GO) term [18] as follow:

$$I(go_n) = log_2 \frac{k(go_n)}{\bigcup_{i=1}^{m} k(go_i)} \tag{7}$$

where go_n represent a GO term, $k(go_n)$ the gene set annotated by go_n and m the number of annotations in the biological process ontology.

2.6   Transcription Factor Classification

To identify human transcription factors, we selected the human genes annotated as "transcription factor activity" in Gene Ontology and the list of transcription factors (TFs) from Transfac Professional [19]. From this list, we removed general TFs (e.g. stable complexes like polymerases or TATA-box-binding proteins), and added some TFs not annotated by GO, producing a final list of 1,722 TFs, from which 632 were on the filtered microarray gene set.

2.7   GeneWays

GeneWays is a computer system designed for automatic analysis of literature data to extract knowledge about molecular interactions [5]. It provides a list of gene pairs associated with a keyword (action), defining the interaction type, and a score between -1 and 1.

2.8   ARACNE

ARACNE is an information-theoretic method for identifying transcriptional interactions between gene products using microarray expression profile data [6]. ARACNE has proven effective in identifying transcriptional targets in complex mammalian networks [15]. We used the bootstrapping version of ARACNE with a list of 632 transcription factors.

2.9   Motifs Enrichment

The combined P-D/P-P interaction network is represented as two independent graphs where the nodes are genes products and a directed or undirected edge represents respectively a P-D or a P-P interaction. Directed edges point from a transcription factor to its target. Randomized networks were built to have the same connectivity properties as the real network. Specifically, the randomized networks have identical distribution for the following properties: (a) P-P interaction degree (number of P-P interactions) per node, (b) P-D interaction in-degree (number of edges pointing to the node) and (c) P-D interaction out-degree (number of edges originating at the node). Randomized networks with this connectivity constraint were built with the igraph library of the statistical

software R. Statistical significance of motif enrichment in the real network was
obtained by computing the zscore:

$$zscore = \frac{N_{real} - mean(N_{random})}{sd(N_{random})} \qquad (8)$$

where N_{real} is the number of motifs in the real network, and $mean(N_{random})$
and $sd(N_{random})$ are the mean and the standard deviation of the number of
motifs in 1,000 randomized networks.

3   Results and Discussion

We used separate Naïve Bayes classifiers to predict P-P and P-D interactions.
This requires positive and negative Gold Standard datasets (GSP and GSN) for
both P-P and P-D interactions to evaluate the likelihood ratio (LR) of each evidence source. Construction of these datasets is described in the previous section.
Note that, as for similar approaches [7,8], we consider fixed P-P and P-D priors
in this paper. This is only a first order approximation and it will need to be adjusted in a Protein and TF-specific way, as additional evidence is collected, since
cellular network connectivity appears to be approximated by a power-law [2,15].

3.1   P-P Interactions

To infer P-P interactions, we integrated the following P-P interaction evidence:
(a) non human interactions for four eukaryotic organisms, (b) two yeast two-
hybrid (Y2H) datasets, (c) the GeneWays literature datamining algorithm [5],
(d) the Gene Ontology biological process annotations [18], and (e) gene co-
expression data from a large collection of 254 B cell expression profiles [15].
Each evidence source was represented as categorical data (continuous values
were binned as necessary) and used to compute a LR based on the GSP and
GSN data as further described (Table 1). Note that we only considered LR
greater than 1 for the different evidences.

**Non human interactions for four eukaryotic organisms and human
Yeast two-hybrid (Y2H):** We extracted putative P-P interactions clues from
IntAct and BIND for the three model organisms *Caenorhabditis elegans*, *Drosophila melanogaster* and *Mus musculus* and from IntAct, BIND and MIPS [20]
for *Saccharomyces cerevisiae*. We defined four different groups of predicted P-P
interactions, one for each organism, by mapping model organisms' genes to human genes using the Inparanoid database that describes eukaryotic orthologous
clusters [21]. As these four sources contain redundant information, we chose to
combine them, together with human interactions extracted from the two Y2H
experiments, in one non-redundant source. In this final group, interactions were
classified according to the number of evidence sources supporting them (from
1 to 5) for computing a LR. As expected, the LR distribution shows that interactions between genes that are supported by more than one data source are

likely to predict P-P interactions between the corresponding orthologous genes in *Homo sapiens*.

GeneWays literature datamining algorithm: By studying the action keyword enrichment for 6,904 P-P interactions in the GSP (from the HPRD), which were also reported by GeneWays, we identified 19 action keywords associated with P-P interactions. These include the following: *assemble, associate, bind, coexpress, coimmunoprecipitate, colocalize, connect, coprecipitate, copurify, dephosphorylate, dissociate from, form, form a complex, immunoprecipitate, interact, recruit, required for, synergize* and *ubiquitinate*. Enrichment was computed with a fisher exact test ($p - value < 10^{-3}$). This list allowed us to extract 25,985 putative GeneWays P-P interactions among 5,797 genes. These were further classified in two groups according to their score ($s \leq 0$ and $s > 0$, respectively). The LR was computed independently for the two groups.

Gene Ontology biological process annotation: It was also observed that interacting proteins tend to share the same biological process [22]. Thus, GO

Table 1. P-P and P-D interaction evidence and Likelihood Ratio (LR)

	Evidence	Bins	LR
	Protein-Protein Interactions	1	33
		≥ 2	848
	GeneWays	≤ 0	165
		> 0	404
	Gene Ontology	< 6	13
		6-7	29
		7-8	39
		8-9	95
		9-10	174
P-P Interaction		10-11	203
Integration		11-12	321
		> 12	496
	Mutual Information	0.22-0.27	2
		0.27-0.32	4
		0.32-0.37	8
		0.37-0.42	22
		0.42-0.47	37
		0.47-0.52	83
		0.52-0.57	127
		0.57-0.62	326
		> 0.62	1713
	Mouse Protein-DNA Interactions		42
P-D Interaction	GeneWays	≤ 0	3
Integration		> 0	10
	ARACNE	< 0.27	3
		≥ 0.27	24

annotations provide additional clues about a P-P interaction. We assembled a list of 4,510,212 human gene pairs sharing a biological process annotation. They were classified using the information content of each GO term, retaining the highest value in case of multiple annotations. This information was binned in 8 groups to compute the LR. The LR distribution shows that GO categories with higher information content, reflecting very precise functional similarity, are more likely to support a P-P interaction than those with smaller values.

Gene co-expression data: It has been established that some interacting proteins, especially those in stable complexes, tend to be co-expressed [23]. Thus co-expression in a large expression profile dataset can provide clues about P-P interactions. We computed mutual information (MI) between 6,083 human genes using their mRNA expression levels measured by the Affymetrix chip HG-U95Av2 in 254 normal, tumor related, and experimentally manipulated B cell populations [15]. We binned the MI into 9 categories to classify the gene pairs. As expected, the LR significantly increases with the MI, reflecting the fact that interacting proteins tend to be co-expressed.

3.2   P-D Interactions

We combined information on P-D interactions from different sources including (a) mouse data from Transfac Professional [19] and BIND and (b) human P-D interactions inferred by the GeneWays and ARACNE algorithms. The data from each clue was binned and tested against the GSP and GSN to compute the LR, reflecting the ability of individual clues to predict MYC targets and, by generalization, other transcriptional interactions (Table 1).

Mouse data from Transfac Professional: We extracted mouse P-D interactions from the Transfac Professional and BIND databases and used the Inparanoid database to predict human P-D interactions, selecting the genes associated to a cluster with a score of 1 only. We defined 551 potential interactions involving 431 genes.

GeneWays: GeneWays interactions contained 250 interactions from Transfac Professional, revealing enrichment for 12 actions: *activate, depend on, include, independence, influence, mediate, regulate, repress, transactivate* and *up-regulate*. In the case where we found enrichment for an action in both P-D interaction and P-P interaction groups (e.g. bind) we retained the action for the group that showed the most significant enrichment for that action. This list allowed us to extract 4,141 potential human P-D interactions, involving 1,754 genes, further classified into two groups according to their score. The LR was computed for the two groups.

ARACNE: ARACNE was successful in predicting MYC targets that were experimentally validated, allowing us to use these results to compute the reliability of ARACNE predictions. We classified the predicted MYC targets according to their MI for computing the LR, revealing that a MYC target with a MI greater than 0.27 has a $p > 50\%$ probability to be a true interaction. We also assumed

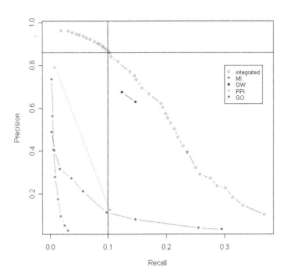

Fig. 1. 10-fold cross-validation: Precision (TP/TP+FP) vs. Recall (TP/TP+FN) curve for the individual and integrated sets (GO = Gene Ontology, GW = GeneWays, MI = Mutual Information, PPI = interactions in model organisms and human Y2H data). TP (True Positive), FP (False Positive) and FN (False Negative) were calculated as GSP and GSN interactions.

that this threshold would produce result similar to those for MYC for other TFs. This was biochemically validated using the BCL6 transcription factor (data not reported). Here, ARACNE predicted 76,251 P-D interactions in human B cells. Note that the MI used for categorizing the LR was computed using the same version as for the new bootstrapping version of ARACNE.

3.3   Bayesian Integration

The Naïve Bayes classifications allowed integrating different sources in a final set of 15,278 P-P interactions (4,373 genes) and 16,640 P-D interactions (462 TFs and 2,026 putative targets) with a posterior probability $p > 50\%$ of being true interactions. We called this set a *mixed interaction* network. The P-P interaction LR distribution (see Table 1) shows that each individual clue is not sufficient to predict interactions, except for clues from strong gene co-expression and from model organisms and Y2H. This last group was expected to be a good predictor as it already intrinsically combines different information sources. Considering P-D interactions, except for 551 interactions predicted from mouse data, only P-D interactions that are ARACNE positive could achieve sufficient LR to exceed the significance threshold ($LR_0 = 21$) determined by the prior (see Table 1). Since ARACNE inferences depend on expression profile data (which is cell-context specific), we claim that the transcriptional part of the network is B cell specific.

To evaluate the performance of the P-P interaction classifier, we computed precision and recall for each evidence source and for the final integrated set, using a ten-fold cross-validation process (Figure 1). For an LR_0 threshold of 800, our network achieves recall of up to 10% with precision always greater than 86%. These measures also illustrate that the interaction clues, when combined together, are a much better predictor than each one taken independently.

In a previous Naïve Bayes classification of human P-P interactions [8], the authors defined 38,379 P-P interactions involving 5,791 genes, using a LR threshold of 381. With this threshold, we identify 40,161 putative P-P interactions among 7,603 genes. Of these, 3,995 are common to the two studies supporting 19,072 P-P interactions in the Rhodes set and 19,226 in our set respectively. Of these, 3,201 were common to both studies, corresponding to 17% of each of the predicted sets. This small, yet highly statistically significant overlap can be explained by the use of our highly context-specific gene expression profile dataset, which is likely to identify interactions that are specific to B cells. We thus consider our P-P interaction interactome to be at least partially indicative of interactions that are B cell specific. Similarly, since the most significant contribution to the total LR comes from the ARACNE algorithm, we also consider P-D interactions to be B cell specific.

3.4   Mixed Interaction Network and Motifs Analysis

To build the final mixed interaction network, we included all missing interactions in the GSPs as well as transcriptional interactions for the TFs reported in Transfac Professional and BIND (respectively 25,473 P-P and 2,798 P-D interactions). The final network contains 40,751 P-P (7,888 genes) and 19,370 P-D (3,768 genes) interactions. Respectively, 12,209 and 16,445 of these were new (i.e., not previously in the GSP or Transfac and BIND). We searched this network for three gene motifs that were highly statistically enriched with respect to the null hypothesis (1,000 randomized networks of identical connectivity). We were particularly interested in two highly enriched regulatory motifs (referred to as R1 and R2) combining both interaction types (Table 2).

Table 2. Regulatory motifs involving P-D and P-P interactions

Motif		#motifs		
Name	Representation	Real Network	Random Network (mean ± SD)	Z-score
R1		23,056	3,496 ± 109	179
R2		3,735	801 ± 153	19

R1 motifs (z-score $Z_{R1} = 179$) describe the regulation of two proteins in a complex by the same TF, suggesting that genes encoding proteins that interact in a complex tend to have a common regulatory program. Among the 6,107 P-P interactions in R1 motifs, 2,037 are jointly regulated by more than one TF (see Supplementary Table S1). These combinatorial regulation events were highly statistically significant in the real network compared to the null hypothesis, highlighting regulation of protein complexes of higher complexity such as for example the ribosome or the collagen. As an illustration, we reported the common targets of CEBPD and MAF that regulate 48 genes encoding proteins organized in complexes (Fig. 2a).

Table 3. Enriched motifs with at least 30% new interactions. P-value was computed with a fisher exact test and reported non-corrected and corrected for multiple testing (bonferroni correction).

TF1	TF2	common targets #targets	%new	z-score	Gene Ontology BP Annotation	P-value (corrected)
CEBPB	CEBPG	3	67	9	–	–
ELK1	ELK3	2	50	14	–	–
IRF8	IRF1	2	50	8	–	–
IRF1	SPI1	3	50	7	antimicrobial humoral response GO:0019735	9.10^{-4} (3.10^{-1})
SMAD4	TFE3	2	50	11	–	–
RB1	RBL1	2	50	13	–	–
SRF	YY1	4	50	3	muscle development GO:0007517	2.10^{-4} (7.10^{-2})
CEBPB	SPI1	4	38	6	antimicrobial humoral response GO:0019735	2.10^{-3} (6.10^{-1})
FOS	SRF	4	38	6	positive regulation of cell proliferation GO:0008284	2.10^{-3} (8.10^{-1})
IRF1	NFKB1	4	38	7	natural killer cell activation GO:0030101	1.10^{-4} (4.10^{-2})
FOS	NFKB1	6	33	9	natural killer cell activation GO:0030101	3.10^{-4} (1.10^{-1})
GATA1	GATA2	3	33	9	–	–

R2 motifs (z-score $Z_{R2} = 19$), on the other hand, show that TFs in a complex tend to regulate the same target genes. Only 597 TF pairs are represented in the 3,735 R2 motifs, indicating that many TF pairs regulate more than one gene. Specifically, 66 TF pairs - with statistically independent expression profiles - were found to regulate two or more common targets (see Supplementary Table S2). We report 12 motifs containing at least 30% new interactions (Table 3). This list shows several TF complexes involving proteins from the same family, such as CEBPB and CEBPG (Fig. 2b), as well as non-related proteins, such as YY1 and SRF, known to bind to the same DNA sites (CArG boxes) on target gene promoters (Fig. 2c). Additionally, some TF complexes included proteins known

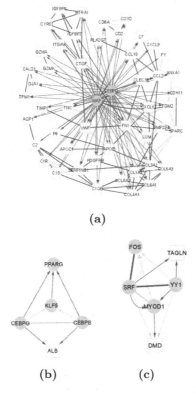

(a)

(b)   (c)

Fig. 2. Examples of enriched regulatory motifs. Undirected and directed edges represent P-P and P-D interactions respectively colored in blue and red if new and in grey if in the GSP. (a) Protein complex regulated by CEBPD and MAF. (b) CEBPB-CEBPG motif. (c) SRF-YY1 motif.

to bind to distinct binding sites. For example, SMAD4 and TFE3 bind respectively to a 4 base pair Smad element and to an E-box. These sites were found to be adjacent in the promoter of known target genes [24]. These differences may help distinguish among different cooperative regulation modes: two modes are associated with either a TF complex binding to a single binding site or two adjacent sites in the promoter of the target genes, while the third mode is associated with two TFs independently binding to different sites on the promoter.

4   Conclusions

The proposed framework constitutes the first example of a mammalian mixed interaction network. Transcriptional cell context specificity was achieved by constraining the inferred P-D interactions on clues dependent on expression profile data. P-P interactions are more likely affected by protein availability than by changes in P-P affinity in different cell types.

5   Supplementary Materials

Supplementary materials are available at:
http://wiki.c2b2.columbia.edu/califanolab/index.php/Publications

Acknowledgments

We thank Andrey Rzhetsky and Raul Rodriguez-Esteban (Columbia University) for providing us GeneWays interactions and for their helpful comments. This work was supported by the National Cancer Institute (R01CA109755), the National Institute of Allergy and Infectious Diseases (R01AI066116), and the National Centers for Biomedical Computing NIH Roadmap Initiative (U54CA121852).

References

1. Yeger-Lotem, E., Sattath, S., Kashtan, N., Itzkovitz, S., Milo, R., Pinter, R.Y., Alon, U., Margalit, H.: Network motifs in integrated cellular networks of transcription-regulation and protein-protein interaction. In: Proc Natl Acad Sci. USA 101(16), 5934–5939 (2004)
2. Yu, H., Xia, Y., Trifonov, V., Gerstein, M.: Design principles of molecular networks revealed by global comparisons and composite motifs. Genome Biol. R7(7), R55 (2006)
3. Rual, J.F., Venkatesan, K., Hao, T., Hirozane-Kishikawa, T., Dricot, A., Li, N., Berriz, G.F., Gibbons, F.D., Dreze, M., Ayivi-Guedehoussou, N., Klitgord, N., Simon, C., Boxem, M., Milstein, S., Rosenberg, J., Goldberg, D.S., Zhang, L.V., Wong, S.L., Franklin, G., Li, S., Albala, J.S., Lim, J., Fraughton, C., Llamosas, E., Cevik, S., Bex, C., Lamesch, P., Sikorski, R.S., Vandenhaute, J., Zoghbi, H.Y., Smolyar, A., Bosak, S., Sequerra, R., Doucette-Stamm, L., Cusick, M.E., Hill, D.E., Roth, F.P., Vidal, M.: Towards a proteome-scale map of the human protein-protein interaction network. Nature 437(7062), 1173–1178 (2005)
4. Stelzl, U., Worm, U., Lalowski, M., Haenig, C., Brembeck, F.H., Goehler, H., Stroedicke, M., Zenkner, M., Schoenherr, A., Koeppen, S., Timm, J., Mintzlaff, S., Abraham, C., Bock, N., Kietzmann, S., Goedde, A., Toksoz, E., Droege, A., Krobitsch, S., Korn, B., Birchmeier, W., Lehrach, H., Wanker, E.E.: A human protein-protein interaction network: a resource for annotating the proteome. Cell 122(6), 957–968 (2005)
5. Rzhetsky, A., Iossifov, I., Koike, T., Krauthammer, M., Kra, P., Morris, M., Yu, H., Duboue, P.A., Weng, W., Wilbur, W.J., Hatzivassiloglou, V., Friedman, C.: Geneways: a system for extracting, analyzing, visualizing, and integrating molecular pathway data. J. Biomed Inform. 37(1), 43–53 (2004)
6. Margolin, A.A., Nemenman, I., Basso, K., Wiggins, C., Stolovitzky, G., Favera, D., Califano, A.: Aracne: An algorithm for the reconstruction of gene regulatory networks in a mammalian cellular context. BMC Bioinformatics 7(Suppl 1), S1–7 (2006)
7. Jansen, R., Yu, H., Greenbaum, D., Kluger, Y., Krogan, N.J., Chung, S., Emili, A., Snyder, M., Greenblatt, J.F., Gerstein, M.: A bayesian networks approach for predicting protein-protein interactions from genomic data. Science 302(5644), 449–453 (2003)

8. Rhodes, D.R., Tomlins, S.A., Varambally, S., Mahavisno, V., Barrette, T., Kalyana-Sundaram, S., Ghosh, D., Pandey, A., Chinnaiyan, A.M.: Probabilistic model of the human protein-protein interaction network. Nat. Biotechnol. 23(8), 951–959 (2005)

9. Peri, S., Navarro, J.D., Amanchy, R., Kristiansen, T.Z., Jonnalagadda, C.K., Surendranath, V., Niranjan, V., Muthusamy, B., Gandhi, T.K., Gronborg, M., Ibarrola, N., Deshpande, N., Shanker, K., Shivashankar, H.N., Rashmi, B.P., Ramya, M.A., Zhao, Z., Chandrika, K.N., Padma, N., Harsha, H.C., Yatish, A.J., Kavitha, M.P., Menezes, M., Choudhury, D.R., Suresh, S., Ghosh, N., Saravana, R., Chandran, S., Krishna, S., Joy, M., Anand, S.K., Madavan, V., Joseph, A., Wong, G.W., Schiemann, W.P., Constantinescu, S.N., Huang, L., Khosravi-Far, R., Steen, H., Tewari, M., Ghaffari, S., Blobe, G.C., Dang, C.V., Garcia, J.G., Pevsner, J., Jensen, O.N., Roepstorff, P., Deshpande, K.S., Chinnaiyan, A.M., Hamosh, A., Chakravarti, A., Pandey, A.: Development of human protein reference database as an initial platform for approaching systems biology in humans. Genome Res. 13(10), 2363–2371 (2003)

10. Hermjakob, H., Montecchi-Palazzi, L., Lewington, C., Mudali, S., Kerrien, S., Orchard, S., Vingron, M., Roechert, B., Roepstorff, P., Valencia, A., Margalit, H., Armstrong, J., Bairoch, A., Cesareni, G., Sherman, D., Apweiler, R.: Intact: an open source molecular interaction database. Nucleic Acids Res. 32(Database issue), D452–D455 (2004)

11. Bader, G.D., Betel, D., Hogue, C.W.: Bind: the biomolecular interaction network database. Nucleic Acids Res. 31(1), 248–250 (2003)

12. Xenarios, I., Salwinski, L., Duan, X.J., Higney, P., Kim, S.M., Eisenberg, D.: Dip, the database of interacting proteins: a research tool for studying cellular networks of protein interactions. Nucleic Acids Res. 30(1), 303–305 (2002)

13. Kumar, A., Agarwal, S., Heyman, J.A., Matson, S., Heidtman, M., Piccirillo, S., Umansky, L., Drawid, A., Jansen, R., Liu, Y., Cheung, K.H., Miller, P., Gerstein, M., Roeder, G.S., Snyder, M.: Subcellular localization of the yeast proteome. Genes Dev. 16(6), 707–719 (2002)

14. Zeller, K.I., Jegga, A.G., Aronow, B.J., O'Donnell, K.A., Dang, C.V.: An integrated database of genes responsive to the myc oncogenic transcription factor: identification of direct genomic targets. Genome Biol. 4(10), R69 (2003)

15. Basso, K., Margolin, A.A., Stolovitzky, G., Klein, U., Dalla-Favera, R., Califano, A.: Reverse engineering of regulatory networks in human b cells. Nat Genet. 37(4), 382–390 (2005)

16. Wang, K., Banerjee, N., Margolin, A., Nemenman, I., Califano, A.: Genome-wide discovery of modulators of transcriptional interactions in human b lymphocytes. In: Apostolico, A., Guerra, C., Istrail, S., Pevzner, P., Waterman, M. (eds.) RECOMB 2006. LNCS (LNBI), vol. 3909, pp. 348–362. Springer, Heidelberg (2006)

17. Alterovitz, G., Xiang, M., Mohan, M., Ramoni, M.F.: Go pad: the gene ontology partition database. Nucleic Acids Res. 35, D322–D327 (2006)

18. Ashburner, M., Ball, C.A., Blake, J.A., Botstein, D., Butler, H., Cherry, J.M., Davis, A.P., Dolinski, K., Dwight, S.S., Eppig, J.T., Harris, M.A., Hill, D.P., Issel-Tarver, L., Kasarskis, A., Lewis, S., Matese, J.C., Richardson, J.E., Ringwald, M., Rubin, G.M., Sherlock, G.: Gene ontology: tool for the unification of biology. The gene ontology consortium. Nat. Genet. 25(1), 25–29 (2000)

19. Matys, V., Fricke, E., Geffers, R., Gossling, E., Haubrock, M., Hehl, R., Hornischer, K., Karas, D., Kel, A.E., Kel-Margoulis, O.V., Kloos, D.U., Land, S., Lewicki-Potapov, B., Michael, H., Munch, R., Reuter, I., Rotert, S., Saxel, H., Scheer, M., Thiele, S., Wingender, E.: Transfac: transcriptional regulation, from patterns to profiles. Nucleic Acids Res. 31(1), 374–378 (2003)

20. Mewes, H.W., Frishman, D., Mayer, K.F., Munsterkotter, M., Noubibou, O., Pagel, P., Rattei, T., Oesterheld, M., Ruepp, A., Stumpflen, V.: Mips: analysis and annotation of proteins from whole genomes in 2005. Nucleic Acids Res. 34(Database issue), D169–D172 (2006)

21. O'Brien, K.P., Remm, M., Sonnhammer, E.L.: Inparanoid: a comprehensive database of eukaryotic orthologs. Nucleic Acids Res. 33(Database issue), D476–D480 (2005)

22. Vazquez, A., Flammini, A., Maritan, A., Vespignani, A.: Global protein function prediction from protein-protein interaction networks. Nat. Biotechnol. 21(6), 697–700 (2003)

23. Ge, H., Liu, Z., Church, G.M., Vidal, M.: Correlation between transcriptome and interactome mapping data from saccharomyces cerevisiae. Nat Genet. 29(4), 482–486 (2001)

24. Hua, X., Miller, Z.A., Wu, G., Shi, Y., Lodish, H.F.: Specificity in transforming growth factor beta-induced transcription of the plasminogen activator inhibitor-1 gene: interactions of promoter dna, transcription factor mue3, and smad proteins. Proc Natl Acad Sci. USA 96(23), 13130–13135 (1999)

Identification and Evaluation of Functional Modules in Gene Co-expression Networks

Jianhua Ruan[1] and Weixiong Zhang[1,2]

[1] Department of Computer Science and Engineering,
[2] Department of Genetics,
Washington University in St. Louis
St. Louis MO 63130, USA
{jruan,zhang}@cse.wustl.edu

Abstract. Identifying gene functional modules is an important step towards elucidating gene functions at a global scale. In this paper, we introduce a simple method to construct gene co-expression networks from microarray data, and then propose an efficient spectral clustering algorithm to identify natural communities, which are relatively densely connected sub-graphs, in the network. To assess the effectiveness of our approach and its advantage over existing methods, we develop a novel method to measure the agreement between the gene communities and the modular structures in other reference networks, including protein-protein interaction networks, transcriptional regulatory networks, and gene networks derived from gene annotations. We evaluate the proposed methods on two large-scale gene expression data in budding yeast and Arabidopsis thaliana. The results show that the clusters identified by our method are functionally more coherent than the clusters from several standard clustering algorithms, such as k-means, self-organizing maps, and spectral clustering, and have high agreement to the modular structures in the reference networks.

Keywords: clustering, community identification, microarray, co-expression networks.

1 Introduction

Many biological sub-systems considered in systems biology can be modeled as networks, where nodes are entities such as genes or proteins, and edges are the relationships between pairs of entities. Examples of biological networks include protein-protein interaction (PPI) networks [1], gene co-expression networks [2], metabolic networks [3], and transcriptional regulatory networks [4]. Much effort has been devoted to the study of their overall topological properties and similarities to other real-world networks [5,6,7,8].

A large amount of available gene expression microarray data has provided opportunities for studying gene functions on a global scale. Since genes that are on the same pathways or in the same functional complex are often regulated by the same transcription factors (TFs), they usually exhibit similar expression

T. Ideker and V. Bafna (Eds.): Syst. Biol. and Comput. Proteomics Ws, LNBI 4532, pp. 57–76, 2007.

patterns under diverse temporal and physiological conditions. Therefore, an important step in analyzing gene functions is to cluster genes according to their expression patterns. The clusters can then be analyzed in several ways. For example, from the promoter sequences of the genes in the same cluster, one may identify common short DNA sequences, which can often suggest the regulation pathways of the genes; in addition, if the majority of the genes in a cluster are known to have some common functions, it is likely that the unannotated genes in the same cluster may also share similar functions. (See [9] for a review). The most popular clustering techniques for gene expression data include hierarchical clustering [10], k-means clustering [11], and self-organizing maps (SOM) [12].

However, genes of similar expression patterns may not necessarily have the same or similar functions. Genes could be accidentally co-regulated or co-expressed [2]; a single event often activate multiple pathways that have distinct biological functions. On the other hand, genes with related functions may not show any close correlation in their expression patterns. For example, there might be time-shift between the expression patterns of genes in the same pathway [13]. Most existing clustering algorithms do not take these possibilities into account.

Another challenging problem for clustering algorithms is to determine the most appropriate number of clusters without prior knowledge of the data. For most clustering algorithms, such as k-means and SOM, it is the user's responsibility to decide the number of clusters to be computed, and it is always possible for the algorithms to return the specified numbers of clusters, regardless of the structure of the data.

To objectively evaluate and validate clustering results is also a daunting task. Generally, different clustering algorithms provide different results and unveil different aspects of the data. To assess the quality of clustering results, most studies have focused on the separation between clusters or homogeneity within clusters [14]. Such numerical evaluation methods depend solely on the data and face a common dilemma: one cannot maximize both the separation and homogeneity at the same time. More importantly, these methods seldomly perform any reality check. For example, does a clustering make any biological sense? Several alternative approaches have been proposed to validate clustering results with biological knowledge, for example, using annotations in the gene ontology (GO) [15,16]. However, these methods are usually affected by factors such as the number of clusters and the distribution of cluster sizes, and cannot precisely measure clustering qualities.

Here, we take a network-based perspective to efficiently *identify* and *evaluate* intrinsic modular structures embedded in large gene expression data. Given the expression profiles of a set of genes, we first construct a co-expression (CoE) network, where the nodes in the network are genes, and the edges reflect expression similarities between pairs of genes. We then apply an algorithm that we have developed recently to identify natural communities in the network, which are densely connected subgraphs that are unexpected by chance [17,18]. Compared to existing clustering methods, our algorithm is relatively independent of any detailed domain knowledge, and can automatically determine the best number of clusters based on the internal structure of the data. Furthermore, we also propose a method to evaluate the biological significance of the clustering results based on their agreement with the structure of other reference biological networks.

We apply the methods to two large gene expression datasets, one for yeast and the other for Arabidopsis. We evaluate the clustering results on yeast genes with three reference networks, including a protein-protein interaction (PPI) network [19], a network based on GO annotations [20], and a network based on TF biding data measured with ChIP-chip technology [21], and the results on Arabidopsis genes with a GO-based reference network. We compare our results with several popular clustering algorithms, including k-means, SOM and spectral clustering, which are applied directly to the expression data. The comparison shows that our network-based approach discovers significantly more enriched functional groups, which also have a better agreement with the reference networks.

The paper is organized as follows. In section 2, we describe the method for constructing gene CoE networks, the algorithm for community identification, and the approach for cluster evaluation. In section 3, we first present some topological results of the CoE networks, then discuss our clustering results and compare them with the results from several popular clustering algorithms. We conclude in section 4 with some discussion.

2   Methods

2.1   Constructing Gene CoE Networks

Many methods have been proposed for constructing CoE networks from gene expression data. The most popular methods first compute a similarity between the expression profiles of every pair of genes, and determine a threshold to select pairs of genes to be connected [22,23,24]. The problem with this type of approaches, aside from being arbitrary in choosing a threshold, is that gene CoE often exhibits a local-scaling property. For example, genes in one cluster may be highly correlated to one another, while genes in another group may be only loosely correlated. Therefore, if we choose a stringent threshold value, many genes in a loosely correlated group may become unconnected. On the other hand, if we attempt to include more gene in the network, the threshold may have to be so low that a large portion of genes are almost completely connected, making further analysis a difficult task. For example, to construct a CoE network for the 3000 yeast genes that we will see in Section 3.1, even if we allow 10% of the genes to be unconnected, the majority of the genes still have more than 300 links (Fig. 1).

We propose a rank-based transformation of similarity matrices to deal with such local-scaling property. We first calculate the Pearson correlation coefficient (or some other similarity measures) between every pair of genes. Then for every gene, we rank all other genes by their correlation coefficients to the gene. Given the ranks, we connect every gene to its top α co-expressed genes, where α is a

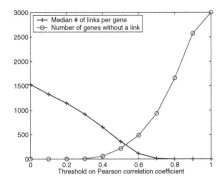

Fig. 1. Median number of CoE links per gene and the number of genes without a CoE link as a function of the threshold on the Pearson correlation coefficient

user defined threshold, with values typically smaller than 5. Note that although the correlation coefficient matrix, C, is symmetric, i.e. $C(i,j) = C(j,i)$, the rank of gene i with respect to gene j, $R(i,j)$, is in general not equal to the rank of gene j with respect to gene i, $R(j,i)$.

This network has several important features. First, all nodes are connected, since each node is connected to at least α other nodes. By varying α, we obtain networks of different granularities. Second, some nodes may have more than α edges, due to the asymmetric property of the ranking. That is, although gene A lists only α genes as its friends, other genes that are not in A's friend list may have A as their friends. In other words, the network can be viewed as directed, even though the directions are ignored in our clustering. In section 3.1, we will show that a CoE network thus constructed has a prominent topological feature different from the CoE networks obtained in previous studies [2,24,25].

A network constructed with this procedure may be different from the underlying biological network that regulates the genes. Nevertheless, at a higher level, the network may capture some topological properties of the actual regulatory network and preserve functional relationships among genes. Genes that are in the same pathway or functional complex tend to be close to one another in the network, i.e., they are often directly linked to each other or connected by short paths. As we will see in section 3, clustering of such networks can indeed produce biologically more meaningful modules than clustering the original expression data with a conventional clustering method. We will also show that clustering of this network is rather robust, in that perturbing a large fraction of its connections does not significantly affect the final clustering results.

2.2 Community Identification

Identifying community structures in a network is similar, but not equivalent, to the conventional graph partitioning problem; both amount to clustering vertices into densely connected subgraphs [26]. A key difference is that for the former, we need to decide whether there are indeed natural communities and how many communities exist in a given network. In contrast, in conventional graph partitioning, the user has to decide how many clusters to look for.

We recently proposed a spectral-based community identification method [17,18]. The method has several unique features. First, it considers local neighborhood information of each node to improves clustering quality [17]. Second, the algorithm combines a modularity function Q to automatically determine the most appropriate number of clusters in a network. Third, the algorithm can handle networks of several thousands of nodes in a few minutes, much faster than most existing algorithms, while often achieving better clustering qualities. We have extensively tested the algorithm on many simulated networks and real-world networks with known community structures, as well as several real applications such as PPI networks and scientific collaboration networks. The results from these analyses show that our method is both efficient and effective. The detailed analysis and evaluation of the algorithm can be found in [18]. Here we briefly describe the key ideas in the algorithm.

Modularity Function. To determine the optimal community structure of a network, Newman and Girvan [27] recently proposed a modularity function, Q, which is defined as:

$$Q(\Gamma_k) = \sum_{i=1}^{k}(e_{ii} - a_i^2), \tag{1}$$

where Γ_k is a clustering that partitions the nodes in a graph into k groups, e_{ii} is the fraction of edges with both nodes within cluster i, and a_i is the fraction of edges with one or both nodes in cluster i. Intuitively, the Q function measures the percentage of edges fully contained within the clusters, subtracted by what one would expect if the edges were randomly placed. The value of Q is between -1 and 1; a larger Q value means stronger community structures. If a partition gives no more within-cluster edges than expected by chance, $Q \leq 0$. For a trivial partitioning with a single cluster, $Q = 0$. It has been observed that most real-world networks have $Q > 0.3$ [28]. The Q function can also be extended to weighted networks straightforwardly by generalizing e_{ii} and a_i to edge weights, instead of number of edges.

It has been shown empirically that higher Q values correspond to better clusters in general [27,29]. Therefore, the Q function provides a good quality measure to compare different community structures, and can serve as an objective function to search for the optimal clustering of a network.

The $Qcut$ Algorithm. Several clustering algorithms have been developed based on *approximate* optimization of Q (as surveyed in [29]), since the optimization is NP-hard [30]. Among them, a spectral algorithm NJW [31], can approximately optimize Q if the number of clusters (k) is given, as shown in [32]. To automatically determine the number of clusters, the NJW algorithm is executed multiple times, with k ranging from the user defined minimum K_{min} to maximum K_{max} number of clusters. The k that gives the highest Q value is deemed the most appropriate number of clusters. The idea has been implemented recently by others and us [32,17].

While this idea is effective in finding community structures in small networks, it scales poorly to large networks, because it needs to execute NJW, whose running time is $O(n^2)$, up to K_{max} times. Without any prior knowledge of a network, one may over-estimate K_{max} in order to reach the optimal Q. In the worst case, K_{max} can be linear in the number of vertices, making it impractical to iterate over all possible k's for large networks.

In order to develop a method that scales well to large networks while retaining effectiveness in finding good communities, we developed an algorithm, called $Qcut$, to recursively divide a network into smaller ones while optimizing Q [18].

Given the adjacency matrix of a network G, we apply the standard NJW spectral clustering algorithm [31] to search for an up to l-way partitioning, where l is a small integer ($l < 5$ typically), that gives the highest Q value. Then, the algorithm is recursively applied to each subnetwork, until the overall Q value cannot be improved by any further partitioning. At each step, a (sub)network is divided into k subnetworks, where k is between 2 and l, and can be different for each partitioning. To reduce computation cost, we restrict l to small integers. We find that with l as small as 3 or 4, the $Qcut$ algorithm can significantly improve the Q values over standard two-way partitioning strategies [33,32], and is much more efficient than direct k-way methods [32,17]. After each split and at the end of all splits, an efficient procedure is applied to fine-tune the clusters in order to further improve the modularity, making $Qcut$ one of the most effective (in terms of accuracy) and efficient algorithms in community identification.

2.3   Cluster Evaluation

A conventional way for evaluating clustering results is to measure separation and homogeneity. We are more interested in the biological soundness and relevance of the clustering results. Therefore, we use two methods based on gene functional annotations to evaluate clustering qualities obtained from gene CoE networks.

Statistical Enrichment of GO Terms. To assess the functional significance of gene clusters, we first compute the enrichment of GO terms for the genes within each cluster. The statistical significance of GO term enrichment is measured by a cumulative hypergeometric test [34]. The p-values are adjusted by Bonferroni corrections for multiple tests [34]. The search of enriched GO terms is performed with a computer program GO::TermFinder [35].

To compare different clustering results, we count the number of GO terms enriched in the clusters at a given significance level. Furthermore, to rule out the possibility that a single cluster may contain a very large number of enriched GO terms and therefore dominate the contribution from other clusters, we also count the number of clusters that have at least one enriched GO term at a given significance level. Note that two clustering results cannot be compared by this method if they differ significantly in numbers of clusters or cluster size distributions, which may strongly affect the number of enriched GO terms. The results of the comparison also depend on what p-value threshold is used.

Evaluation Using Reference Networks. We propose a novel method for assessing clustering qualities based on external information of the genes. The basic idea is to introduce a functional reference network (discussed later), and compare the clustering of the CoE network with the reference networks. In such a reference network, genes are linked by edges that represent certain functional relationships between them, where the edges may be weighted according to the reliability or significance of the relationships. This network can be expected to have some modular structures as well. Since our purpose is to identify functional modules within a CoE network, we would prefer a good clustering of the CoE network to represent a good partitioning of the reference network as well; i.e., genes within the same CoE clusters should be connected by many high weight edges in the reference network, while genes in different CoE clusters should share less functions or be connected with low weight edges in the reference network. To measure the agreement between the clustering of a CoE network and a reference network, we force the reference network to be partitioned exactly the same way as the CoE network, i.e., the group memberships of the nodes in the reference network are the same as that of the CoE network. We then compute the modularity of the reference network using Equation (1). Since the modularity score is not biased by the number of clusters or the cluster size distributions, it can be applied to compare arbitrary clustering results.

Now that we have introduced the measurement, what can be a reference network and how do we get it? First, many available biological networks, such as PPI networks and genetic interaction networks, can be adopted directly. Evidently, however, some networks may be more suitable than others for evaluating gene CoE clusters.

In general, a reference network does not have to be directly observed from experiments, but rather derived from knowledge about the genes. Two genes can be connected if they posses some common attributes or features, given that the common attributes are related to CoE. For example, they may participate in the same biological process or be regulated by a common TF. These types of information can be represented by a matrix, where each row is a gene, and each column is an attribute. To construct a network from the matrix, genes are treated as nodes, and an edge is drawn between two genes if they share at least one common attribute. Edges are weighted by some similarity measure of genes' attributes. To measure the similarity, we use a well-developed function in document clustering that takes into account the significance of attributes [36]. For example, the GO terms GO:0009987 (cellular process), which is very close to the root of the GO graph and has a large number of genes associated, is not very informative in clustering genes and should be weighted less than the GO term GO:0045911 (positive regulation of DNA recombination).

Denote a gene-attribute matrix by $A = (a_{ij})$, where $a_{ij} = 1$ if gene i has attribute j, or 0 otherwise. A is transformed into a weighted matrix $W = (w_{ij})$, where $w_{ij} = a_{ij} \times idf_j$. The weighting factor idf_j, called the inverse document frequency (IDF) [36], is defined by $idf_j = \log(n/\sum_i a_{ij})$, where n is the number of genes. With this transformation, the attributes that occur in many genes receive low weights in W. The edge weight between two genes is then measured by the cosine of their weighted attribute vectors:

$$S_{ij} = cos(w_{i.}, w_{j.}) = \frac{\sum_k w_{ik} w_{jk}}{\sqrt{\sum_k w_{ik}^2 \sum_k w_{jk}^2}}, \tag{2}$$

where $w_{i.}$ and $w_{j.}$ are the i-th and j-th rows of W, respectively. As expected, many genes may be connected with very low weights if they share some non-specific functions. We apply a weight cutoff to remove such edges. We have found, however, that the result is almost not affected by the use of different cutoff values, as discussed in Results section.

We use three types of reference networks to evaluate clusters. The first is a network constructed from biological process GO annotations [20], with each term as an attribute. The ontology and annotation files for yeast and Arabidopsis genes are downloaded from http://www.geneontology.org/. To construct a reference network, we first convert the original annotation files to include complete annotations, i.e., if a gene is associated with a certain term, we also add all ancestors of the term into the gene's attribute list due to term inheritance. If two terms are associated with exactly the same set of genes, we remove one to avoid double counting. We also remove GO terms that are associated with more than 500 or less than 5 genes. The procedure results in 1034 and 438 GO terms for yeast and Arabidopsis, respectively. The second is a PPI network for budding yeast, downloaded from the BioGRID database [19]. We combined all physical interactions obtained from yeast two-hybrid or affinity purification-mass spectrometry experiments. The edges are weighted by the number of times an interaction was observed. The third network is a co-binding network derived from the ChIP-chip data of 203 yeast transcription factors (TFs) under rich media conditions [21]. We treat each TF as an attribute, and construct a network with the procedure described above. We only consider a binding as genuine if its p-value is less than 0.001, according to the original authors [21].

3   Results

3.1   Topology of Yeast CoE Networks

Previous studies have analyzed the topologies of various networks, including biological and social networks, and suggested a common scale-free property [5,6,7,8]. In a scale-free network, the probability for a node to have n edges obeys a power-law distribution, i.e. $P(n) = c \times n^{-\gamma}$. The implication of the scale-free property is that a few nodes in the network are highly connected, acting as hubs, while most nodes have low degrees. In contrast, in a random network, connections are spread almost uniformly across all nodes. Real networks also differ from random networks in that the former often have stronger modular structures, reflected by higher clustering coefficients [28].

In this study, we obtained a set of yeast gene expression data measured in 173 different time points under various stress conditions [37], and selected 3000 genes that showed the most expression variations. We constructed four CoE networks

Table 1. Statistics of yeast CoE networks

α	2	3	4	5
# of nodes	3000	3000	3000	3000
# of edges	5432	8103	10775	13432
k_{avg}	3.6	5.4	7.2	9.0
c	0.089	0.124	0.144	0.159
c_r	0.010	0.015	0.018	0.02
c_{sf}	0.002	0.003	0.004	0.005

k_{avg}: averge node degree; c: clustering coefficient; c_r: clustering coefficient of the network constructed from permuted expression data; c_{sf}: clustering coefficient of the rewired network.

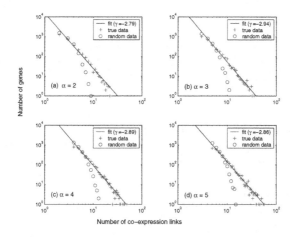

Fig. 2. Distribution of the number of CoE links. Y-axes show the number of genes with a certain number of CoE links (X-axes) in a network.

with $\alpha = 2$, 3, 4 and 5, respectively, i.e., we let each gene connect to its top α correlated genes (see section 2.1). To compare, we also randomly shuffled the real gene expression data, and constructed four networks from the random data with the same α values.

To determine the topological characteristics of the CoE networks, we first plotted the number of genes having n connections as a function of n in a log-log scale. As shown in Fig. 2, the networks constructed from the real data exhibit a power-law degree distribution for all the α values considered, indicating that an overall scale-free topology is a fairly robust feature of the CoE networks. In contrast, the networks constructed from the randomized expression data are close to random networks and contain significantly fewer high-degree nodes. Second, we computed the clustering coefficients of the networks derived from real and randomized expression data. As shown in Table 1, the true CoE networks have much higher clustering coefficients than the random network. Furthermore, we permuted the CoE networks through random rewiring [38], which preserves

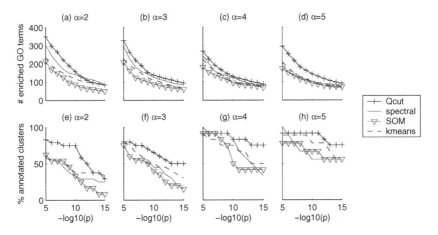

Fig. 3. Enrichment of GO terms in yeast CoE networks. Y-axes in (a)-(d): number of GO terms enriched in the clusters. Y-axes in (e)-(h): percentage of clusters that are enriched with at least one GO term. X-axes: p-value cutoff to consider a GO term enriched.

degree for each node, and thus does not change the scale-free property of the networks. As shown in Table 1, the clustering coefficients of the rewired networks are significantly lower than that of the original networks, indicating that high clustering coefficients is indeed a property of CoE networks.

It is not surprising to see that CoE network is yet another example of scale-free networks. However, several previous studies on a number of gene CoE networks have suggested that there might exist profound topological differences between gene CoE networks and other biological networks [2,23,25]. In these studies, it has been observed that the exponent γ for the power law degree distribution of CoE networks is consistently less than 2, while in other biological networks, including PPI networks and metabolic networks, as well as in real-world social and technology networks, γ is usually between 2 and 3 (for examples see [28,38]). A scale-free network with $\gamma < 2$ has no finite mean degree when its size grows to infinity, and is dominated by nodes with large degrees [28]. To determine the values of γ for the CoE networks that we have constructed, we fitted a linear regression model to each log-log plot to calculate its slope. As shown in Fig. 2, the values of γ in our networks are consistently between 2 and 3, similar to many real-world or biological networks.

The difference in γ between previous CoE networks and ours is most likely due to the difference in the network construction procedures. We used a rank-based method in selecting CoE links, while most existing methods are threshold-based. A threshold-based network tends to include a large number of high degree nodes, and therefore usually have a small γ value. Although further work is required, the similarity in γ values between our networks and other biological and real-world networks suggests that the networks constructed by our method may better represent the underlying functional structures than previous CoE networks.

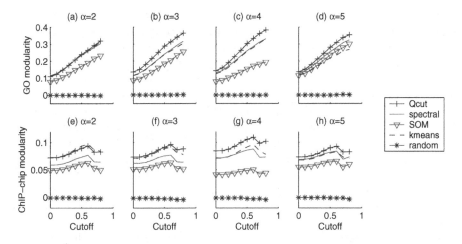

Fig. 4. Agreement between modular structures in yeast CoE networks and two reference networks derived from GO annotations (a-d) and ChIP-chip data (e-h). X-axes: edge weight cutoff for the reference networks.

3.2   Functional Modules in Yeast CoE Networks

We applied the $Qcut$ algorithm to cluster the four CoE networks constructed in section 3.1. The best numbers of clusters suggested by $Qcut$ for the four networks are 24, 20, 12 and 12, respectively. For comparison, we also applied three popular clustering algorithms, including k-means, SOM, and spectral clustering, to the expression data, using Pearson correlation-coefficient as the distance measure. We obtained $k = 24$, 20, 12 and 9 clusters for each of the three competing algorithms. The SOM algorithm was executed on 4×6, 4×5, 3×4, and 3×3 grids to produce the desired number of clusters [12]. Because $Qcut$ identified 12 clusters on both the $\alpha = 4$ and $\alpha = 5$ networks, we matched the 12 clusters of the $\alpha = 5$ network with the 9 clusters from the competing algorithms to avoid redundant comparison. Another reason for this matching is that $Qcut$ often produce a few small clusters, while the clusters of the competing algorithms are relatively uniform in sizes. Therefore, the "effective" number of clusters is smaller for $Qcut$ than for other algorithms, so we used the last test to compensate some differences in the cluster size distributions.

To validate the biological significance of the clusters, we first counted the number of GO terms enriched in the clusters and the number of clusters that had at least one enriched GO term at various significance levels. As shown in Fig. 3, the clusters identified by $Qcut$ contain more enriched GO terms than the competing algorithms for most p-value cutoff levels and for different number of clusters (Fig. 3(a)-(d)). Furthermore, the percentages of clusters containing at least one enriched GO term are also higher for $Qcut$ than for the other algorithms (Fig. 3(e)-(h)). However, as observable from the figure, the number of enriched GO terms increase with the number of clusters. Therefore, it is hard to conclude which network has produced the best clustering result.

Second, we evaluated the clusters with three reference networks that capture different functional interactions between genes: a co-function network based on GO annotations, a co-binding network based on ChIP-chip data, and a PPI network (Section 2.3).

The comparison with all three reference networks indicates that the clusters identified by $Qcut$ have higher agreement with the reference networks than do the clusters by the competing algorithms (Fig. 4 and 5). The spectral clustering algorithm generally performs better than the other two, which is reasonable since the spectral method is able to capture some topological features embedded in the data. We also randomly shuffled the clustering results of $Qcut$ while fixing the sizes of the clusters, and compared the random clusters with the three reference networks. The modularity is always very close to zero (Fig. 4 and 5), meaning that the agreement between our clustering results and the three reference networks is not due to chance.

Among the three reference networks, the GO-based network has higher agreement with the CoE network modules ($Q > 0.35$) than do the PPI network ($Q \approx 0.15$) and the co-binding network ($Q \approx 0.1$). The low agreement between CoE and PPI networks may be partially due to the high level of noises in PPI data. On the other hand, the low agreement between the CoE and co-binding networks is somewhat unexpected, because co-binding should be a relatively strong evidence of CoE. The reason might be that the gene expression data were measured under stress conditions while the ChIP-chip experiments were conducted under normal conditions. Therefore, genes bound by common TFs under normal conditions may not necessarily exhibit similar expression patterns under these stress conditions, and some co-binding under stress conditions were not captured by the ChIP-chip experiments.

For the GO-based reference network, the modularity value is a monotonic increasing function of edge cutoffs, indicating that genes sharing many functions or several specific functional terms are more likely to be co-expressed than genes sharing some broad functional terms. In comparison, the ChIP-chip modularity reaches its peak at cutoff = 0.6, probably because there are relatively fewer genes sharing exactly the same regulators, and therefore the co-binding network becomes very sparse when the cutoff is greater than 0.6. However, the relative performance of different clustering algorithms is not affected by the cutoffs.

Both Fig. 4 and Fig. 5 show that the modules in the $\alpha = 2$ CoE network have the worst agreement with any of the three reference networks, which means that this network might be too sparse to capture all functional relationships. The $\alpha = 4$ CoE network has the highest agreement with the three reference networks, while the networks with $\alpha = 3$ or 5 give slightly worse results.

Furthermore, to test if the competing algorithms may give the best results with a different number of clusters, we applied the spectral clustering to obtain $k = 5, 6, \ldots, 25$ clusters, and computed their agreement to the GO-based reference network at cutoff value = 0.8. As shown in Fig. 6, the spectral clustering achieved best modularity 0.323 at $k = 13$, which is significantly lower than the best modularity of $Qcut$ ($Q = 0.384$).

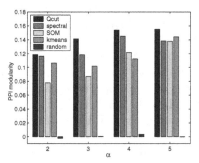

Fig. 5. Agreement between modular structures in yeast CoE networks and a PPI network. The four groups represent four networks with different values of α.

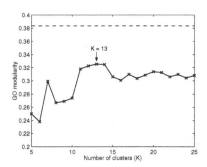

Fig. 6. Agreement between the clusters identified by spectral clustering and the GO-based reference network as a function of the number of clusters (k). The dashed line represents the best agreement achieved by $Qcut$ ($\alpha=4$, k=12).

Finally, Table 2 shows the number of genes within each cluster identified from the $\alpha = 4$ network, the most significantly enriched GO biological process terms, and the transcription factors that may bind to the genes within each cluster. As shown, most clusters contain highly coherent functional groups, and are regulated by a few common transcription factors, e.g., clusters 8, 9, 11 and 12. The majority of the genes in cluster 12 are involved in protein biosynthesis, and can be bound by FHL1 and RAP1, both of which are known to be involved in rRNA processing and regulating ribosomal proteins [39]. Cluster 9 is significantly enriched by genes that are involved in generation of precursor metabolites and energy, and can be bound by HAP4, a TF regulating carbohydrate metabolism [39]. Cluster 2 contains almost two third of the ribosome biogenesis genes, although no TFs bind to this set of genes specifically. Cluster 11 are enriched with genes that can be bound by eight different TFs. Interestingly, these TFs are all known cell-cycle regulators [39].

Several small clusters correspond to very specific functional groups. For example, 17 of 22 genes in cluster 10 are involved in Ty element transposition;

9 of 18 genes in cluster 3 are related to chromatin assembly or disassembly. Six genes in cluster 3 are regulated by HIR1/2/3, which are known to be involved in the transcription of histone genes [39].

Among the 25 genes in cluster 4, 4 genes have a common function in telomere maintenance, while 16 genes encode hypothetic proteins and have unknown functions. Interestingly, 5 of the 16 uncharacterized genes are located near telomeric region [39]. Moreover, A significant number of genes in this cluster are regulated by four common transcription factors (Table 2). Therefore, it is very likely that these uncharacterized genes are closely related to the function or maintenance of telomere. Clusters 5 and 7 contain both a large fraction of genes with unknown functions, and groups of genes with significantly enriched common functions or common TFs. It is possible that these uncharacterized genes also have similar functions to the other annotated genes in the same cluster.

3.3   Robustness of Clustering Results

Since gene expression measurement is inherently noisy, and our method only used the top-ranked CoE edges in network construction, we need to evaluate whether the resulting clusters were stable with respect to perturbations. To this end, we removed all the top three CoE links from the yeast $\alpha = 6$ network. That is, each gene was connected only to its fourth, fifth and sixth best correlated genes. This network has about the same number of edges as the $\alpha = 3$ network, but very different edges. In fact, the edges in the two networks are completely different. To compare their modular structures, we calculated a minimal Wallace Index [40] between the clustering results on the two networks, which is a defined by $W(\Gamma, \Gamma') = \min(N_{11}/S(\Gamma), N_{11}/S(\Gamma'))$, where Γ and Γ' are two clustering results for comparison, N_{11} is the number of pairs of genes in the same cluster in both Γ and Γ', and $S(\Gamma)$ is the number of pairs of genes in the same cluster in Γ.

Surprisingly, the clustering on these two network are fairly similar: the Wallace Index between the two clusters is 0.63, i.e., 63% of the gene pairs are conserved between the two clustering results. In contrast, we would only expect the two clusters to share $(12\pm0.1)\%$ of the gene pairs if the two networks were not related. Furthermore, the clusters obtained from the reduced $\alpha = 6$ network still contain significantly more enriched GO terms than the clusters identified by k-means and SOM (data not shown).

3.4   Functional Modules in an Arabidopsis CoE Network

To test our method on higher organisms, we applied it to a set of Arabidopsis gene expression data downloaded from the AtGenExpress database(`http://www.uni-tuebingen.de/plantphys/AFGN/atgenex.htm`). The dataset contains the expression of 22k Arabidopsis genes in root and shoot in 6 time points following cold stress treatment. We selected the genes that are up- or down-regulated by at least five-folds in at least one of the 6 time points in root or

Table 2. Functional modules in a yeast CoE network

Cluster	Size	Category[1]	Term	Count	Enrichment[2]	P-value
1	361	BP	protein catabolism	32	4.2	2.0E-12
		BP	protein folding	21	5.9	1.6E-11
2	498	BP	ribosome biogenesis	133	9.2	2.0E-106
3	18	BP	chromatin assembly or disassembly	9	36.4	5.3E-13
		TF	HIR2	6	129.8	2.3E-12
		TF	HIR1	6	62.9	3.0E-10
		TF	HIR3	6	57.7	5.3E-10
4	25	BP	telomerase-independent telomere maintenance	4	82.3	1.1E-07
		BP	biological process unknown	16	2.9	7.6E-06
		TF	GAT3	13	56.8	3.5E-21
		TF	YAP5	15	43.5	5.8E-17
		TF	PDR1	9	25.8	3.1E-11
		TF	MSN4	8	35.0	3.8E-11
5	422	BP	spore wall assembly	16	7.0	1.6E-10
		BP	biological process unknown	138	1.5	1.2E-07
		TF	NRG1	21	4.2	1.4E-08
		TF	SUM1	16	3.9	2.3E-06
		TF	PHD1	15	3.4	3.2E-05
6	99	–	–	–	–	–
7	463	BP	carbohydrate metabolism	41	2.9	4.6E-10
		BP	biological process unknown	178	1.7	9.5E-17
		BP	response to stimulus	62	1.7	2.0E-05
		TF	UME6	25	2.5	2.6E-05
		TF	NRG1	15	2.8	3.6E-04
8	108	BP	nitrogen compound metabolism	25	7.0	5.2E-15
		TF	MET31	4	9.6	8.0E-04
		TF	MET32	5	5.7	2.1E-03
9	192	BP	generation of precursor metabolites and energy	50	8.2	7.5E-33
		TF	HAP4	22	9.2	5.1E-16
10	22	BP	Ty element transposition	17	58.6	6.2E-29
		TF	SUM1	4	18.9	5.8E-05
11	604	BP	carboxylic acid metabolism	76	3.0	2.4E-19
		BP	cell organization and biogenesis	212	1.6	3.7E-15
		TF	SWI6	45	2.9	3.6E-11
		TF	SWI4	44	2.8	2.7E-10
		TF	FKH2	35	3.0	4.7E-09
		TF	MBP1	36	2.8	1.9E-08
		TF	STE12	22	3.6	7.9E-08
		TF	NDD1	30	2.9	1.1E-07
		TF	FKH1	34	2.5	9.6E-07
		TF	MCM1	22	2.9	3.9E-06
12	186	BP	protein biosynthesis	131	6.4	6.4E-85
		TF	FHL1	96	17.1	3.3E-105
		TF	RAP1	58	11.5	2.2E-48

[1]For each cluster, significantly enriched biological process GO terms (BP) or binding of transcription factors (TF) are counted.

[2]Fold of enrichment is calcuated as:
$$\frac{(\text{number of genes in cluster with the term}) \times (\text{number of genes in genome})}{(\text{number of genes in cluster}) \times (\text{number of genes in genome with the term})}.$$

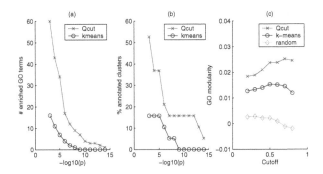

Fig. 7. Enrichment of GO terms in the Arabidopsis CoE network. (a) number of enriched GO terms; (b) percentage of clusters with at least one enriched GO term; (c) agreement between modular structures in the Arabidopsis CoE network and a reference network derived from GO annotations. X-axes in (a) and (b) are p-value cutoff to consider a GO term enriched. X-axis in (c) is edge weight cutoff for the reference network.

shoot. We then constructed a CoE network by connecting each gene to its top three correlated genes (i.e. $\alpha = 3$). The network has 2545 genes and 5838 CoE links.

Our clustering algorithm partitioned the network into 19 clusters, with a Q value of 0.81, indicating strong modular structures. As in the previous experiments, we examined the GO terms enriched in the clusters at various significance levels, and compared them with the results of the standard k-means algorithm that partitions the gene expression data into 19 clusters. As shown in Fig. 7(a) and (b), the clusters identified by our network-based method contains significantly more enriched terms than that identified by the k-means, and GO terms are enriched in more clusters in our method than in k-means. Furthermore, the comparison with a reference network derived from GO annotations (section 2.3) shows that the clusters identified by $Qcut$ is more consistent with the reference network (Fig 7(c)). Note that due to the high complexity of gene expression regulation and the lack of detailed gene annotations, the modularity of the GO network in Arabidopsis is much lower than that of yeast (0.025 vs 0.38).

Table 3 shows the most enriched functional categories for each cluster. Some clusters are enriched with functions that are known to be related to cold stress responses, e.g. clusters 7 (photosynthesis), 11 (circadian rhythm), 14 (response to heat), 15 (antiporter activity) and 18 (lipid binding). Since the annotation for the Arabidopsis genome is much poorer than that for the yeast genome, the enrichment of GO terms in the clusters for Arabidopsis genes are not as significant as that for the yeast genes. On the other hand, our method may be applied to assign putative functional roles to some of these unannotated genes.

Table 3. Functional modules in an Arabidopsis CoE network

Cluster	Size	GO term	Count	Enrichment*	P-value
1	199	-	-	-	-
2	141	-	-	-	-
3	79	-	-	-	-
4	180	catalytic activity	99	1.6	4.1E-09
		amino acid and derivative metabolism	18	4.4	3.9E-08
5	284	endomembrane system	79	1.6	3.5E-06
6	238	oxidoreductase activity	40	2.6	7.7E-09
		secondary metabolism	18	3.1	9.9E-06
7	65	photosynthesis	11	32.6	8.7E-16
8	261	RNA binding	11	4.6	9.2E-06
9	186	galactolipid biosynthesis	3	17.6	1.8E-04
10	19	branched-chain-amino-acid transaminase activity	3	172.6	1.7E-07
11	117	starch metabolism	4	16.0	5.0E-05
		circadian rhythm	6	7.6	8.5E-05
12	271	protein modification	37	2.1	4.3E-06
13	268	methyltransferase activity	8	4.7	1.4E-04
14	13	response to heat	8	87.7	1.9E-15
15	223	antiporter activity	10	6.1	1.5E-06
16	151	transcription regulator activity	60	3.0	2.5E-17
17	200	zeaxanthin epoxidase activity	3	16.4	2.2E-04
18	17	lipid binding	5	48.2	2.9E-08
		membrane	12	2.7	1.8E-04
19	249	calcium ion binding	13	3.2	1.1E-04

*See Table 2

4   Conclusions and Discussion

In this paper, we proposed a network-based method for clustering microarray gene expression data, and a method for evaluating clustering results based on reference networks. We introduced a simple rank-based method to construct gene CoE networks from microarray data, and applied a spectral clustering algorithm that we developed recently to cluster networks into densely connected sub-graphs. We applied our method to two gene expression datasets, and showed that the network-based clustering method can produce biologically more meaningful clusters than conventional methods such as k-means and SOM. The clusters identified by our methods contain significantly more enriched GO terms than other algorithms and exhibited better agreement with several reference networks.

It is rather surprising that the simple method we proposed to construct CoE networks worked well. The connections in such a CoE network are obviously different from actual biological interactions. Nevertheless, at a higher level, the CoE networks that we constructed have captured most topological properties and functional relationships in the true network. We expect that a more sophisticated method for constructing CoE networks, such as Bayesian networks [41] and Boolean networks [42], may improve the discovery of function modules even further.

The CoE networks that we constructed posses a unique topological feature that is different from the CoE networks reported in the literature. In our network, the exponent of the power-law degree distribution falls in the range of 2 to 3, similar to most other real-world networks, whereas the exponent of CoE networks reported in the literature is below the critical value of 2. We are currently looking for the causes of this discrepancy and examining their effects on our clustering algorithm.

Although we have only demonstrated our method on gene expression data, it can be applied to other types of experimental data as well. The efficiency of our clustering method and its relative independence of any detailed domain knowledge of the data make it well suited for identifying intrinsic structures in large-scale network data. Furthermore, the cluster evaluation method we proposed may be used as a general framework for assessing different algorithms and comparing clustering results based on external knowledge.

Acknowledgements

This research was supported in part by an NSF grants ITR/EIA-0113618 and IIS-0535257, and a grant from Monsanto Company.

References

1. Tong, A., Drees, B., Nardelli, G., Bader, G., Brannetti, B., Castagnoli, L., Evangelista, M., Ferracuti, S., Nelson, B., Paoluzi, S., Quondam, M., Zucconi, A., Hogue, C., Fields, S., Boone, C., Cesareni, G.: A combined experimental and computational strategy to define protein interaction networks for peptide recognition modules. Science 295, 321–324 (2002)
2. Stuart, J., Segal, E., Koller, D., Kim, S.: A gene-coexpression network for global discovery of conserved genetic modules. Science 302, 249–255 (2003)
3. Jeong, H., Tombor, B., Albert, R., Oltvai, Z., Barabasi, A.: The large-scale organization of metabolic networks. Nature 407, 651–654 (2000)
4. Lee, T., Rinaldi, N., Robert, F., Odom, D., Bar-Joseph, Z., Gerber, G., Hannett, N., Harbison, C., Thompson, C., Simon, I., Zeitlinger, J., Jennings, E., Murray, H., Gordon, D., Ren, B., Wyrick, J., Tagne, J., Volkert, T., Fraenkel, E., Gifford, D., Young, R.: Transcriptional regulatory networks in saccharomyces cerevisiae. Science 298, 799–804 (2002)
5. Jeong, H., Mason, S., Barabasi, A., Oltvai, Z.: Lethality and centrality in protein networks. Nature 411, 41–42 (2001)
6. Ravasz, E., Somera, A., Mongru, D., Oltvai, Z., Barabasi, A.: Hierarchical organization of modularity in metabolic networks. Science 297, 1551–1555 (2002)
7. Barabasi, A., Oltvai, Z.: Network biology: understanding the cell's functional organization. Nat. Rev. Genet 5, 101–113 (2004)
8. Oltvai, Z., Barabasi, A.: Systems biology. life's complexity pyramid. Science 298, 763–764 (2002)

9. Armstrong, N., van de Wiel, M.: Microarray data analysis: from hypotheses to conclusions using gene expression data. Cell Oncol. 26, 279–290 (2004)

10. Eisen, M., Spellman, P., Brown, P., Botstein, D.: Cluster analysis and display of genome-wide expression patterns. Proc. Natl. Acad. Sci. USA 95, 14863–14868 (1998)

11. Tavazoie, S., Hughes, J., Campbell, M., Cho, R., Church, G.: Systematic determination of genetic network architecture. Nat. Genet. 22, 281–285 (1999)

12. Tamayo, P., Slonim, D., Mesirov, J., Zhu, Q., Kitareewan, S., Dmitrovsky, E., Lander, E., Golub, T.: Interpreting patterns of gene expression with self-organizing maps: methods and application to hematopoietic differentiation. Proc. Natl. Acad. Sci. USA 96, 2907–2912 (1999)

13. Qian, J., Dolled-Filhart, M., Lin, J., Yu, H., Gerstein, M.: Beyond synexpression relationships: local clustering of time-shifted and inverted gene expression profiles identifies new, biologically relevant interactions. J. Mol. Bio. 314, 1053–1066 (2001)

14. Bolshakova, N., Azuaje, F.: Machaon CVE: cluster validation for gene expression data. Bioinformatics 19, 2494–2495 (2003)

15. Azuaje, F., Al-Shahrour, F., Dopazo, J.: Ontology-driven approaches to analyzing data in functional genomics. Methods Mol. Biol. 316, 67–86 (2006)

16. Gibbons, F., Roth, F.: Judging the quality of gene expression-based clustering methods using gene annotation. Genome Res. 12, 1574–1581 (2002)

17. Ruan, J., Zhang, W.: Identification and evaluation of weak community structures in networks. In: Proc. National Conf. on AI, (AAAI-06). pp. 470–475 (2006)

18. Ruan, J., Zhang, W.: Discovering weak community structures in large biological networks. Technical Report cse-2006-20, Washington University in St Louis (2006)

19. Stark, C., Breitkreutz, B., Reguly, T., Boucher, L., Breitkreutz, A., Tyers, M.: Biogrid: a general repository for interaction datasets. Nucleic Acids Res. 34, D535–539 (2006)

20. The Gene Ontology Consortium: The gene ontology (GO) database and informatics resource. Nucleic Acids Res. vol. 32 (2004)

21. Harbison, C., Gordon, D., Lee, T., Rinaldi, N., Macisaac, K., Danford, T., Hannett, N., Tagne, J., Reynolds, D., Yoo, J., Jennings, E., Zeitlinger, J., Pokholok, D., Kellis, M., Rolfe, P., Takusagawa, K., Lander, E., Gifford, D., Fraenkel, E., Young, R.: Transcriptional regulatory code of a eukaryotic genome. Nature. 431, 99–104 (2004)

22. Zhou, X., Kao, M., Wong, W.: Transitive functional annotation by shortest-path analysis of gene expression data. Proc Natl Acad Sci. USA 99, 12783–12788 (2002)

23. Carter, S., Brechbuhler, C., Griffin, M., Bond, A.: Gene co-expression network topology provides a framework for molecular characterization of cellular state. Bioinformatics 20, 2242–2250 (2004)

24. Zhu, D., Hero, A., Cheng, H., Khanna, R., Swaroop, A.: Network constrained clustering for gene microarray data. Bioinformatics 21, 4014–4020 (2005)

25. Aggarwal, A., Guo, D., Hoshida, Y., Yuen, S., Chu, K., So, S., Boussioutas, A., Chen, X., Bowtell, D., Aburatani, H., Leung, S., Tan, P.: Topological and functional discovery in a gene coexpression meta-network of gastric cancer. Cancer Res. 66, 232–241 (2006)

26. Fjallstrom, P.: Algorithms for graph partitioning: A survey. Linkoping Electron. Atricles in Comput. and Inform. Sci. (1998)

27. Newman, M., Girvan, M.: Finding and evaluating community structure in networks. Phys Rev. E. Stat Nonlin Soft Matter Phys 69, 26113 (2004)

28. Newman, M.: The structure and function of complex networks. SIAM Review 45, 167–256 (2003)

29. Danon, L., Duch, J., Diaz-Guilera, A., Arenas, A.: Comparing community structure identification. J. Stat. Mech, p. P09008 (2005)
30. Garey, M., Johnson, D.: Computers and Intractability: A Guide to the Theory of NP-completeness. Freeman, San Francisco (1979)
31. Ng, A.Y., Jordan, M.I., Weiss, Y.: On spectral clustering: Analysis and an algorithm. In: NIPS. pp. 849–856 (2001)
32. White, S., Smyth, P.: A spectral clustering approach to finding communities in graph. In: SIAM Data Mining (2005)
33. Shi, J., Malik, J.: Normalized cuts and image segmentation. IEEE Trans. Pattern Anal. Mach. Intell. 22, 888–905 (2000)
34. Altman, D.: Practical Statistics for Medical Research. Chapman & Hall/CRC (1991)
35. Boyle, E., Weng, S., Gollub, J., Jin, H., Botstein, D., Cherry, J., Sherlock, G.: Go:termfinder - open source software for accessing gene ontology information and finding significantly enriched gene ontology terms associated with a list of genes. Bioinformatics 20, 3710–3715 (2004)
36. Jones, K.S.: Idf term weighting and ir research lessons. Journal of Documentation 60, 521–523 (2004)
37. Gasch, A., Spellman, P., Kao, C., Carmel-Harel, O., Eisen, M., Storz, G., Botstein, D., Brown, P.: Genomic expression programs in the response of yeast cells to environmental changes. Mol. Biol. Cell 11, 4241–4257 (2000)
38. Albert, R., Barabasi, A.: Statistical mechanics of complex networks. Reviews of Modern Physics 74, 47 (2002)
39. Saccharomyces genome database. http://www.yeastgenome.org/
40. Wallace, D.L.: Comment. Journal of the American Statistical Assocation 78, 569–576 (1983)
41. Friedman, N., Linial, M., Nachman, I., Peer, D.: Using bayesian networks to analyze expression data. J. Comput Biol. 7, 601–620 (2000)
42. Kauffman, S.: A proposal for using the ensemble approach to understand genetic regulatory networks. J. Theor Biol. 230, 581–590 (2004)

A Linear Discrete Dynamic System Model for Temporal Gene Interaction and Regulatory Network Influence in Response to Bioethanol Conversion Inhibitor HMF for Ethanologenic Yeast

Mingzhou (Joe) Song[1] and Z. Lewis Liu[2]

[1] Department of Computer Science, New Mexico State University
P.O. Box 30001, MSC CS, Las Cruces NM 88003, U.S.A
[2] National Center for Agricultural Utilization Research
U.S. Department of Agriculture, Agriculture Research Service
1815 N University Street, Peoria, Illinois 61604, U.S.A

Abstract. A linear discrete dynamic system model is constructed to represent the temporal interactions among significantly expressed genes in response to bioethanol conversion inhibitor 5-hydroxymethylfurfural for ethanologenic yeast *Saccharomyces cerevisiae*. This study identifies the most significant linear difference equations for each gene in a network. A log-time domain interpolation addresses the non-uniform sampling issue typically observed in a time course experimental design. This system model also insures its power stability under the normal condition in the absence of the inhibitor. The statistically significant system model, estimated from time course gene expression measurements during the earlier exposure to 5-hydroxymethylfurfural, reveals known transcriptional regulations as well as potential significant genes involved in detoxification for bioethanol conversion by yeast.

1 Introduction

Computational modeling of gene regulatory networks (GRNs) is a central focus in systems biology. By far, few approaches are capable of describing the information flow over time in a large network. Only a dynamic system model of a GRN can empower biologists to fully understand the interactions among entities in a network. Verhulst equation, a discrete dynamic system model of one variable, is an example that is widely used in mathematical biology (Edelstein-Keshet, 2004) to study population dynamics in evolution. Although early work that utilizes difference equations to model GRNs exists (D'haeseleer et al., 1999), which estimates system coefficients by least squares, the potential of discrete dynamic systems in modeling GRNs has remained largely unrecognized until recent endeavors by systems biology researchers such as Bonneau et al. (2006) and Schlitt and Brazma (2006), who characterize gene interactions by discrete dynamic system models composed of linear difference equations or finite state linear

T. Ideker and V. Bafna (Eds.): Syst. Biol. and Comput. Proteomics Ws, LNBI 4532, pp. 77–95, 2007.
© Springer-Verlag Berlin Heidelberg 2007

equations. Our work moves along with three innovations. The first is to perform log-time domain interpolation to reposition non-uniformly spaced samples to equally spaced time locations. The second is to assess statistical significance of all possible linear difference equations for a given gene node and to choose the most significant one, as well as to assess the statistical significance of the entire system. The third is to enforce power stability on the discrete dynamic system model so that it does not exhibit chaotic or unstable behaviors under a normal condition. A discrete dynamic system is power stable if variables in the system stay bounded as time goes to infinity given a bounded initial state.

A major motivation of our work originates from the investigation of genetic mechanisms for bioethanol conversions in yeast in pursuit of renewable sources of energy. As interest in alternative energy sources rises, the concept of agriculture as an energy producer has become increasingly attractive. Renewable biomass, including lignocellulosic materials and agricultural residues, has become attractive low cost materials for bioethanol production. One major barrier of biomass conversion to ethanol is inhibitory compounds generated during biomass pretreatment, which interfere with microbial growth and subsequent fermentation. For economic reasons, dilute acid hydrolysis is commonly used to prepare the biomass degradation for enzymatic saccharification and fermentation (Bothast and Saha, 1997; Saha, 2003). However, numerous side-products are generated by this pre-treatment, many of which inhibit microbial metabolism. More than 100 compounds have been detected to have potential inhibitory effects on microbial fermentation (Luo et al., 2002). Among these compounds, 5-hydroxymethylfurfural (HMF) and furfural are the most potent and representative inhibitors derived from biomass pretreatment (Taherzadeh et al., 2000; Martin and Jonsson, 2003). Other commonly recognized inhibitors include acetic acid, cinnamic acid, coniferyl aldehyde, ethanol, ferulic acid, formic acid, levulinc acid, and phenolics. Few yeast strains tolerant to inhibitors are available due to a lack of understanding of mechanisms involved in the stress tolerance for bioethanol fermentation. Based on functional genomic studies, a concept of genomic adaptation to the biomass conversion inhibitors by the ethanologenic yeast is proposed (Liu and Slininger, 2006a; Liu, 2006). However, a great deal of detailed knowledge of GRNs involved remains unknown.

In the computational and biological context described above, we have developed discrete dynamic system models to study the genetic basis underlying metabolic pathway of the ethanologenic yeast. As initiated in this study, we have delineated through discrete dynamic system models how a biological system behaves in response to inhibitor HMF during the earlier exposure to the inhibitor for ethanol production. In this model, the change in expression level of a target gene at a discrete time point is a linear function of the expression levels of influential genes at previous discrete time points. This model facilitates the characterization of gene interactions in efficient production of ethanol in yeast under both control and stress conditions, allowing one to introduce specific perturbations into a system and predict the effects on biomass conversion under various stress conditions. Furthermore, the model enables one to identify relevant genes and gene interactions for optimal genetic manipulations that will guide the engineering of more robust yeast strains for economic ethanol production.

Although other alternative modeling methodologies have been developed, discrete dynamic system models are advantageous given the increased availability of experimental designs that collect time-course gene expressions at the whole system scale. *Dynamic Bayesian networks* (DBNs) extend the static Bayesian networks by introducing the time aspect. Both models have been used for modeling GRNs: the former used by (Ong et al., 2002) and the latter used by (Imoto et al., 2003; Friedman, 2004). A DBN describes statistical dependencies among genes temporally, by extending Bayesian networks to incorporate time transitions between the Bayesian networks at consecutive time points. Since a DBN does not describe functional relations among genes, it is not a suitable tool to understand the dynamics of a GRN, though there is no doubt that Bayesian networks and DBNs are indeed successful in extracting probabilistic dependencies among genes. The *Boolean networks* (Liang et al., 1998; Akutsu et al., 2003; Pal et al., 2005) have gained momentum recently. Shmulevich et al. (2002) introduce stochastic components for GRNs by creating probabilistic Boolean networks. Since a Boolean network represents gene expression level in two states: on and off, this qualitative abstraction limits its capacity in discriminating quantitative changes in gene expression levels under perturbed situations. Our primary goal is to establish a gene interaction network model inferring regulatory mechanisms in biomass conversion to ethanol, especially the quantitative shift of biotransformation and detoxification of the inhibitors, which requires information beyond the presence or absence of genes. Thus, Boolean networks are not the best dynamic strategy to describe accurately the amount of ethanol product as a function of the concentration of glucose substrate. *Differential equations* in both deterministic (Meir et al., 2002) and stochastic (van Kampen, 1997) formulations have been used to model interactions among entities in a GRN in continuous time. The E-CELL Project (Tomita et al., 1999; Takahashi, 2004; Takahashi et al., 2005) targets at reproducing *in silico* intracellular biochemical and molecular interactions within a single cell with the differential equation model. The stochastic differential equations (Master equations) represent the dynamics of probabilities of states by differential equations, which is impractical for GRNs involving more than a handful of genes because the amount of data needed to characterize stochastic behaviors is subject to curse of dimension, to be encountered in probability density estimation. However, almost all differential equations reduce to difference equations in practical applications. Direct discrete dynamic modeling overrides this intermediate step and speaks the native discrete time language of a computer. We believe it is more effective to go without the intermediate mode of differential equations. In addition, the time interval between discrete points in difference equations can be adjusted to the sparsity of data, making it more flexible to model the dynamics at different resolutions.

2   Results and Discussion

Using first order linear difference equations, we build discrete dynamic system models for the transcriptional interactions among genes in yeast during the earlier exposure to the inhibitor HMF for ethanol production. In a discrete dynamic system model, the expression change rate of a gene is a linear function of the concentrations of potential regulator genes – one equation is used for each gene. A network is derived from a discrete dyamic system model by creating an edge from every potential regulator to each gene it regulates. These models were developed based on mRNA abundance over five time points in the presence or absence of HMF. Data were collected with two biological replications each with two technical replications.

An inferred interaction network with a subset of 46 gene nodes plus an HMF node is depicted in Fig. 1. Based on ANOVA and cluster analysis, 46 significantly induced expressed genes by the HMF treatment were selected and used for the prototype computation modeling development. This network model captured temporal dependencies among the 46 genes and HMF during the earlier exposure to the inhibitor in yeast fermentation process. The system model underlying the network is an optimal solution after searching all possible directed graphs with 47 nodes, except that the HMF node is not allowed to have incoming edges and the maximum number of incoming edges for a gene node is at most 5. Existence of an edge from YAP1 to DDI1 indicates a temporal dependency of the rate of change in DDI1 expression on the mRNA level of YAP1. The number 1.2e-07, positioned next to the edge, is the p-value of this temporal dependency. The original system matrix was stabilized by scaling all eigenvalues by the spectral norm 3.09. The overall p-value, 1.6e-5, of the entire system model indicates that the model is statistically significant. The p-value is based on a stringent standard and the resulting model has high levels of consistency with biological observations because the probability of the model arising by chance is as low as 1.2e-07.

Among three known transcription factors, PDR1, PDR3, and YAP1, in this subset of genes, YAP1 was shown as one of the most influential regulators as demonstrated by this model in earlier response to the HMF stress for ethanol production (Fig. 1). This is strongly supported by current knowledge and documented experimental observations (Teixeira et al., 2006). For example, the following edges have been reported as transcriptional regulations including YAP1 to DDI1 (Haugen et al., 2004), YAP1 to ATM1 (Haugen et al., 2004), YAP1 to GRE2 (Lee et al., 1999), YAP1 to SNQ2 (Lee et al., 2002; Lucau-Danila et al., 2005), and YAP1 to TPO1 (Lucau-Danila et al., 2005). Four more edges from YAP1 demonstrated enhancement to SCS7, PDR1, PDR11, and HIS3, suggesting regulatory rules of YAP1 to these genes (Fig. 1). According to YEASTRACT, SCS7, PDR1, PDR11, and HIS3 are considered potential transcriptional regulatees of YAP1 based on sequence motifs (YEA, 2006). In addition, transcriptional factor PDR3 showed regulatory rule to RSB1 as demonstrated in this model, which is in agreement with and supported by previous documented observations (Devaux et al., 2002). It also showed enhancement to SAM3, ATM1,

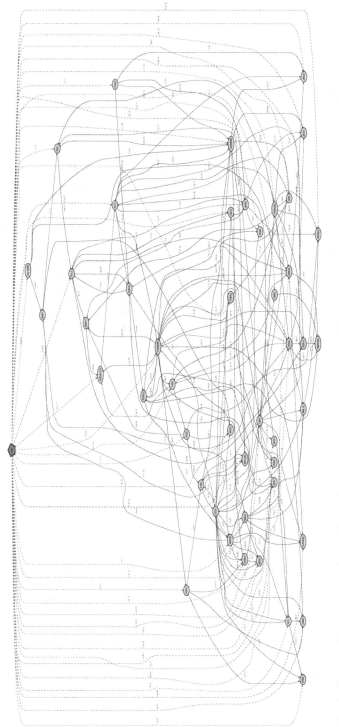

Fig. 1. Temporal interactions for a subset of 46 genes in response to HMF for biomass conversion to ethanol by ethanologenic yeast. The p-values of each edge are displayed. A solid directed edge in green from the first gene node to the second gene node with an arrowhead indicates enhancement of the second gene by the first gene; An edge in red from the first gene node to the second gene node with a solid dot indicates repression of the second gene by the first gene. The dashed edges represent the external influence from HMF to each gene: red for repressing and green for enhancing. The graph is rendered by the software GraphViz (available from www.graphviz.org).

and PDR12. It is very encouraging that the gene regulatory network model developed in this study is highly consistent with the current knowledge including documented experimental observation and sequence motif based analysis. More significantly, the model demonstrated in this study showed statistical significance on the temporal dependencies.

This system model also presented numerous interesting network interactions among genes with potential significance. For example, STE6, SNQ2, ARG4 and YOR1 significantly enhanced directly or indirectly 15, 8, 5, and 4 other genes, respectively. These genes have been observed to be core stress response genes and many related genes are observed to be interested to cope with the HMF stress for survival. Resolution of such interactions could have a significant impact to understand the mechanism of detoxification and the stress tolerance caused by HMF. Although they have not been reported, such statistically significant gene interactions presented by this model could be potentially biologically significant to predict unknown gene interaction networks. With the high consistency between the model network on YAP1 presented in this study and current knowledge, it is reasonable to assume potential relationships presented in this model with significant p-values. However, a common transcription factor PDR1 did not show significant regulatory rule to the selected subset genes in this model. We need to examine it further using biological experiment. Although it is highly homologous with PDR3, PDR1 does not always respond the same with PDR3.

Another impact of the system model is to prescribe desired system behaviors by applying perturbation to the system. A perturbation can be changing the concentration level of the inhibitor HMF, silencing of a subset of genes in the network, or mutating of a subset of genes. To increase the tolerance to the inhibitor HMF, one can consider adjusting the influential genes to achieve an effect similar to the transcriptome profile observed in the absence of HMF. In Fig. 1, the following genes were identified as potential significant elements in gene interaction networks for detoxification and HMF stress tolerance: STE6 (15/46), YCR061W (14/46), YAP1 (12/46), YGR035C (10/46), SNQ2 (8/46), HSP10 (7/46), and YAR066W (7/46). By perturbing these major regulators, one will exert the most control over expressions of other genes, which might be economically desirable.

Another strategy to genetic engineering for wild type yeast to become tolerant to the inhibitor HMF is to study the system model of HMF resistant yeast strains. Preliminary tolerant strains for *in situ* detoxification of the inhibitors have been developed (Liu et al., 2004; Liu and Slininger, 2005; Liu et al., 2005; Palmqvist et al., 1999; Wahbom and Hahn-Hägerdal, 2002). By comparing the system models for the wild type and the tolerant yeast strain, one can identify those genes that behave differently between the two strains. Those different genes can be the targets of genetic engineering for the wild type strains to become HMF tolerant.

Figures 2 to 3 show how well the model fits the observed trajectory data from the 46 genes. The model is able to capture trends in the data precisely such as ARG1, ICY1, MDS3, TPO1, and YCR061W.

Figure 4 demonstrates the prediction made by the model how the time courses evolve differently when the same sample is subject to different experimental conditions.

Figures 2 to 3, each corresponding to a different sample, show how well the model fits the observed data from the 46 genes. In these figures, the original time course sample, the log-time interpolated data, and the fitted time course by the model are illustrated. The model captured the trend in the data precisely for genes such as ARG1, ICY1, MDS3, TPO1, and YCR061W, given the large sample variation present in most microarray experiments. We are primarily interested in detecting significant interactions that can be captured by the capability of linear discrete dynamic system model. The poor fits suggest that there might be nonlinear interactions in addition to the linear interactions, which we plan to address in the future work.

Based on the estimated coefficients in above tables, simulations are performed to evaluate the effectiveness of the estimated difference equations. Figure 4 demonstrates the prediction made by the model how the time courses evolve differently when the same sample is subject to different experimental conditions. It can be observed that the influential gene nodes in Fig. 1 evidently exhibit sharper transitions in the time course than the non-influential genes. The presence of HMF has significantly influenced all the selected genes. However, the effect takes on different courses. Some genes have been enhanced such as ICT1, while staying on similar curvatures; some genes are repressed severely such as TPO1; other genes show opposite transitions such as HIS3.

A strong temporal dependency of gene X on gene Y can indicate a transcriptional regulation from Y to X. However, a real transcriptional regulation from transcription factor W to Z may not show up as a temporal dependency of Z on W due to other factors involved in the expression of W. It is possible that W can have a high concentration of mRNA, but somehow the translation of W mRNA to its protein product is blocked by the presence of other regulatory proteins during translation. Therefore, the mRNA concentration of Z will be low due to the scarcity of the protein product of its transcription factor W. No temporal dependencies of Z on W can be possibly established in such a scenario. It is also plausible that a temporal dependency does not equate to a real transcriptional regulation: Two genes S and T can co-express in similar patterns and only one of them S is a real transcription factor of a third gene R. Although it is unlikely that two genes have identical expression patterns as the nature of biology tends to go parsimony, the measurement may not discern a difference that is below the noise level, which can be high in current microarray technology. These limitations can be overcome when proteome measurements are available and uncertainty in measurements is reduced. They are not inherent problems of discrete dynamic system models.

The methodology presented in this paper can be applied to the analysis of a network from data sets that contain both transcriptome and proteome measured simultaneously on the same sample. With such data sets that encapsulate complete snapshots of molecular processes during bioethanol conversion, the

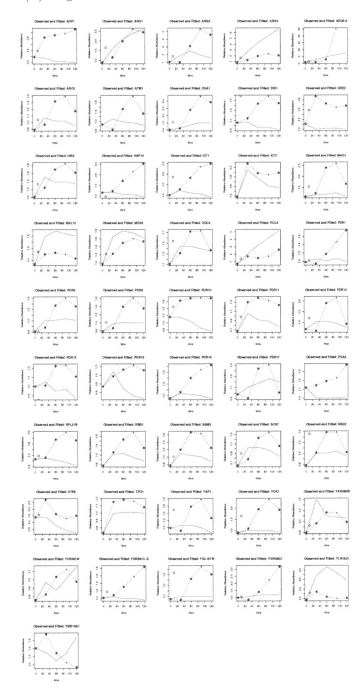

Fig. 2. Sample 1. Control: Not exposed to HMF. Fitted gene expression time courses (green solid lines) from the model versus the observed ones (blue dotted lines). The big open blue circles represent the original values; the small solid blue circles are interpolated values actually used.

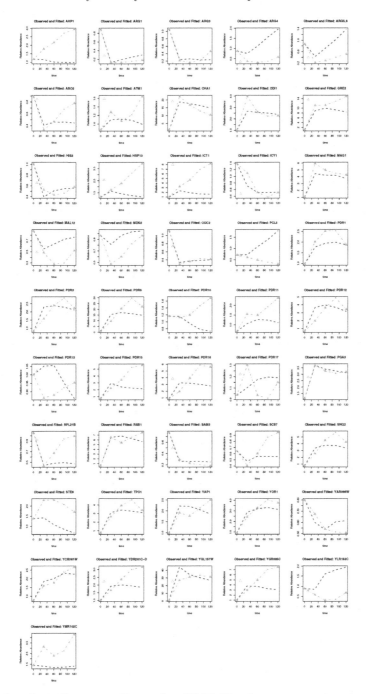

Fig. 3. Sample 2. Treatment: Exposed to HMF. Fitted gene expression time courses (red dashed lines) from the model versus the observed ones (yellow dash-dotted lines). The big open yellow triangles represent the original values; the small filled yellow triangles are interpolated values for model estimation.

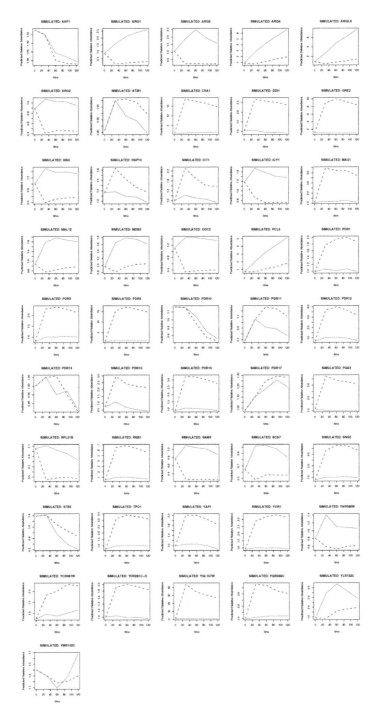

Fig. 4. Predictions of mRNA expression time courses of the 46 genes: HMF in absence (green solid lines) versus HMF in presence (red dashed lines)

temporal dependencies depicted by our approach will be able to provide a more accurate account of the genomic mechanism on inhibitor detoxification and tolerance in ethanologenic yeast.

3   Materials and Methods

3.1   Microarray Experiments, Data Preprocessing, and Gene Expression Analysis

Target genome microarray of *Saccharomyces cerevisiae* was fabricated using GeneMachine OmniGrid 300 microarrayer robot. A recent version of 70-mer oligo set representing 6,388 genes was applied and Codelink activated slides were used. DNA oligo samples were resuspended in 150 mM of sodium phosphate printing buffer (pH 8.5) at a final concentration of 20 M probes for printing. Each genome microarray was designed with two replications on one slide. Each microarray slide consisted of 13,000 elements including target genes and spiking-in quality controls for linear dynamic calibration, ratio reference, DNA sequence background, and slide background controls. The first developed universal external RNA control was applied in microarray experiments (Liu and Slininger, 2006b). The universal quality control consisted of six unique RNA transcripts that can be applied to different assay platforms of microarray and real time quantitative RT-PCR, including SYBR Green and TaqMan probe-based chemistry. It was demonstrated that the signal intensity detected from these controls are independent from cell treatment of stress or environmental conditions in a host RNA background. Highly fitted linearity and dynamic ranges provided a basis for estimation of mRNA abundance in gene expression analysis. Such external RNA controls provide an unbiased normalization reference, valid dynamic range of linearity, and estimate of variations of microarray experiments. It guards reliability and reproducibility of expression data and also makes it possible to compare data derived from different experiments and different assay platforms for data verification and confirmation. Each of the control elements in each array had 48 replications and was distributed evenly in each block of the microarray. A mini array consisting of the controls and two background controls was designed on top of a target genome array with 16 replications. This mini array served as a reference to adjust PMT Gain balance of GenePix 4000B scanner for two dye channels prior to a scanning of the entire target array for data acquisition.

Ethanologenic yeast *S. cerevisiae* NRRL Y-12632 was used and maintained lyophilized in the ARS Culture Collection, National Center for Agricultural Utilization Research, USDA, Peoria, IL. Yeast cultures were incubated on a synthetic complete medium for 6h prior to a treatment by HMF (30 mM) as previously described (Liu et al., 2004). Briefly, HMF was added to the medium in a fleaker fermentation system at 30°C. A set of gene expression profiles derived from a yeast culture grown under the same conditions without the HMF treatment served as a control. The time point of inhibitors added was designated as hour 0. Yeast cells were harvested periodically starting from 0h, 10 min, 30 min, 2h and 4h. Cell samples were harvested by centrifugation at 25°C and immediately frozen and stored at $-80°C$ until use. Total RNA was isolated using a protocol based on Schmitt et al. (1990) with modifications. The RNA was further purified using a nucleic acid purification column. RNA probe was labeled using an indirect dUTP Cy3 or Cy5 labeling procedure based on Hegde et al. (2000) with modifications. Microarray slide was scanned and data acquisition obtained using GenePix 4000B scanner and GenePix Pro software. Pre-scan control was used to adjust PMT Gain against Cy3 and Cy5 channels and ratios of signal intensities between Cy3 and Cy5 were balanced to 1 using the calibration controls determined using the mini-array. Microarray data were analyzed using Gene-Spring program. Control gene CtrlGm_5 was used as normalization reference for each gene. Median of foreground signal intensity subtracted by background for each dye channel was used. Data were filtered between each dye channel and among multiple microarray experiments. A gene list shared by all microarray experiments was generated and used for data analysis. ANOVA was performed to identify genes significantly expressed in comparison with the control. Based on expression patterns, subsets of gene lists were generated by self-organizing map and cluster analysis.

3.2   Log-Time Interpolation

Non-uniform time sampling is often used in a time course experimental design, such that high frequency components in the original continuous signal can be preserved. Conversely, interpolation in the original time domain over non-uniform samples tends to distort high frequency components in the original signal. To save sharp transitions at densely sampled time locations, we apply a logarithm transform on time by

$$t' = \log(t + t_0)$$

where t' is the time variable in the log-time domain. Selection of the constant t_0 is determined by how well it equalizes the distance between each consecutive pair of time points after the log-time transform. The observed samples are then interpolated by cubic splines in the log-time domain, by assuming that the sampling times are designed sufficiently well to capture major change of the signals; or equivalently, the change of gene expression levels between two consecutive time points can be captured by the cubic splines. Let $x = f(t')$ be the interpolated cubic spline. One can obtain values at equally spaced time points $0, h, 2h, \ldots,$ $kh, \ldots,$ in the original time domain by

$$x_k = f(\log(kh + t_0))$$

where h is the sampling interval. We pick the same number of interpolated points as the number of points in the original data set. So the interpolation solely serves to equalize the non-uniform time points in the log-time domain. If more points were interpolated, the p-value must be adjusted to that effect, otherwise, faulty significance might arise. The discrete dynamic system model will be fitted to the interpolated values in the original time domain.

3.3   First-Order Linear Discrete Dynamic System Model

Although dynamics in molecular processes are largely nonlinear – reflected by various nonlinear kinetics models, the number of observations sufficient to induce a nonlinear model for a biological system is too large to be practical for a system with more than a handful of variables. Instead of nonlinear models, we use the first-order linear discrete dynamic model to capture the linear effect of a system. A system can only be considered linear when the perturbation to the system is sufficiently small. A large perturbation could lead the system out to another state of linearity. In our experiment, the time points we collected reflected the initial response of gene expressions to the inhibitor HMF before major dramatic dynamic effect takes place. We consider the linear discrete dynamic system model can approximate primary expression response to HMF.

In a first-order linear discrete dynamic system model, the transition from one state at discrete time t to the next state at $t + 1$ depends linearly on the state of the system at time t. Let h be the constant time span of 1 unit of discrete time. First order refers to the transition from t to $t + 1$ does not depend on the state of the system at $t - 1$, $t - 2$, and so on, except the state at t. Let $\mathbf{g}[t] = [g_1[t], g_2[t], \ldots, g_N[t]]^T$ be a vector of the expression levels of N genes at time t. Let $\mathbf{e}[t] = [e_1[t], e_2[t], \ldots, e_K[t]]^T$ be a vector of the strength of K external signals at time t. A first-order linear discrete dynamic system model can be written as

$$\mathbf{g}[t + 1] - \mathbf{g}[t] = h \ \{A \ \mathbf{g}[t] + B \ \mathbf{e}[t]\} + \epsilon[t] \tag{1}$$

where $A = \{a_{i,j}\}$ is an $N \times N$ system matrix and $a_{i,j}$ ($i \neq j$) is the influence of gene j on gene i, $a_{i,i}$ is the self-control rate, $B = \{b_{i,k}\}$ is an $N \times K$ influence matrix where $b_{i,k}$ is the influence of the k-th signal on gene i, $\epsilon[t] = [\epsilon_1[t], \epsilon_2[t], \ldots, \epsilon_N[t]]^T$ is a vector of noise levels to each gene at time t. The noise is estimated by fitting the linear discrete dynamic system model, and thus is a function of the time interval as well as the observed data. In the modeling process, we assume the noise model Gaussian. We also introduce a possible intercept vector I to the right hand side of the above equation during model selection for each node.

Solving the Linear Difference Equations. From the experiments under different conditions, one can collect M time course observations or trajectories of the system at the discrete time points $0, 1, 2, \ldots, T$. Let $\mathbf{g}^m[0], \mathbf{g}^m[1], \ldots, \mathbf{g}^m[T]$ ($m = 1 \ldots M$) be all the observed system states, and $\mathbf{e}^m[0], \mathbf{e}^m[1], \ldots, \mathbf{e}^m[T]$ be all the external stimulus applied to the system. We use the least squares to find optimal estimates of system matrix A and influence matrix B. The system model defined in Eq. (1) can be written as a collection of all M observations by

$$\mathbf{g}^m[t + 1] - \mathbf{g}^m[t] = h \ \{A \ \mathbf{g}^m[t] + B \ \mathbf{e}^m[t]\} + \epsilon^m[t]$$

where

$$\mathbf{g}^m[t] = \begin{pmatrix} g_1^m[t] \\ g_2^m[t] \\ \vdots \\ g_N^m[t] \end{pmatrix}, \quad \mathbf{e}^m[t] = \begin{pmatrix} e_1^m[t] \\ e_2^m[t] \\ \vdots \\ e_K^m[t] \end{pmatrix}, \quad \text{Noise: } \epsilon^m[t] = \begin{pmatrix} \epsilon_1^m[t] \\ \epsilon_2^m[t] \\ \vdots \\ \epsilon_N^m[t] \end{pmatrix}$$

The above formulation can be rearranged into a multiple linear regression form

$$\mathbf{g}^m[t+1] = (hA + I)\mathbf{g}^m[t] + hB\mathbf{e}^m[t] + \epsilon^m[t]$$

Equivalently, for each gene node, we have

$$g_i^m[t+1] = \left[\sum_{j=1}^{N}(ha_{ij} + I(i=j))g_j^m[t]\right] + \left[\sum_{k=1}^{K} hb_{ik}e_k^m[t]\right] + \epsilon_i^m[t]$$

Let $\mathbf{a}_i = (a_{i1}, \ldots, a_{ik})^\top$ and $\mathbf{b}_i = (b_{i1}, \ldots, b_{ik})^\top$ be the parameters associated with gene node i. Thus, $\mathbf{a}_i$ and $\mathbf{b}_i$ can be solved independently of the other nodes. By least squares, optimal estimates for $\mathbf{a}_i$ and $\mathbf{b}_i$ are

$$\mathbf{b}_i = \left[2h^2 \sum_{m=1}^{M}\sum_{t=1}^{T-1} \mathbf{e}^m[t]\mathbf{e}^m[t]^\top\right.$$

$$- \left(2h^2 \sum_{m=1}^{M}\sum_{t=1}^{T-1} \mathbf{e}^m[t]\mathbf{g}^m[t]^\top\right)\left(2h^2 \sum_{m=1}^{M}\sum_{t=1}^{T-1} \mathbf{g}^m[t]\mathbf{g}^m[t]^\top\right)^{-1}\left(2h^2 \sum_{m=1}^{M}\sum_{t=1}^{T-1} \mathbf{g}^m[t]\mathbf{e}^m[t]^\top\right)\right]^{-1}$$

$$\left[\left(2h \sum_{m=1}^{M}\sum_{t=1}^{T-1} (g_i^m[t+1] - g_i^m[t])\mathbf{e}^m[t]\right)\right.$$

$$- \left.\left(2h^2 \sum_{m=1}^{M}\sum_{t=1}^{T-1} \mathbf{e}^m[t]\mathbf{g}^m[t]^\top\right)\left(2h^2 \sum_{m=1}^{M}\sum_{t=1}^{T-1} \mathbf{g}^m[t]\mathbf{g}^m[t]^\top\right)^{-1}\left(2h \sum_{m=1}^{M}\sum_{t=1}^{T-1} (g_i^m[t+1] - g_i^m[t])\mathbf{g}^m[t]\right)\right]$$

$$(2)$$

$$\mathbf{a}_i = \left(2h^2 \sum_{m=1}^{M}\sum_{t=1}^{T-1} \mathbf{g}^m[t]\mathbf{g}^m[t]^\top\right)^{-1}\left[\left(2h \sum_{m=1}^{M}\sum_{t=1}^{T-1} (g_i^m[t+1] - g_i^m[t])\mathbf{g}^m[t]\right) - \left(2h^2 \sum_{m=1}^{M}\sum_{t=1}^{T-1} \mathbf{g}^m[t]\mathbf{e}^m[t]^\top\right)\mathbf{b}_i\right]$$

$$(3)$$

Selecting the Most Significant Linear Difference Equation for Each Gene Node. For each gene node, the more variables involved in the difference equation for that node, the better the fit. However, statistical significance starts to drop once a maximal complexity has reached to a point that the sample does not support more variables to be involved. Thus, we select the best subset of potential regulators for each gene node such that the corresponding linear difference equation yields the most statistically significant fit. The statistical significance is determined by the F-test. For N genes and K external signals, there are 2^{N+K-1} possible subsets to consider, which is computationally feasible only for a network with less than a dozen of nodes. We limit the number of possible incoming edges or potential regulators for each node to some computational doable number. Although this lead to an incomplete exploration of the system search space, our experience indicates that major influential gene nodes can be identified even when the number of regulator nodes explored is small.

Stabilization. Although solutions to the linear difference equations constitute an optimal fit to the observed data, the resulted system can be unstable, meaning that the log expression levels of some genes increase to infinity or decrease to negative infinity as time goes on when the initial state of the system is finite. Thus we stabilize the system model when no external stimuli are present.

Now we derive the stabilization formula. Equivalently, Eq. (1) can be written as

$$\mathbf{g}[t+1] = (hA + I)\mathbf{g}[t] + hB\,\mathbf{e}[t] + \epsilon[t] \qquad (4)$$

When the system is not subject to external stimulus or noise, it becomes

$$\mathbf{g}[t+1] = (hA + I)\mathbf{g}[t] \qquad (5)$$

In the bioethanol conversion process, this system equation describes the ideal behavior of the yeast gene expression without the inhibitor HMF in a zero-noise environment. In such a system, one does not expect the expression of any gene becomes unstable during the experiment since otherwise the subject perishes. An optimal solution found for A by Eq. (3) might lead to an unstable system in Eq. (5). Let $W = hA + I$. A necessary and sufficient condition for the system described by Eq. (5) to be stable is to require W to be power stable – all eigenvalues of W must be located within or on the unit circle; or the spectral norm must be no greater than one. Let $\lambda(W)$ be the sequence of eigenvalues of W. The spectral norm $\rho(W)$ is defined by (Golub and van Loan, 1996)

$$\rho(W) = \max\{|\lambda| : \lambda \in \lambda(W)\}$$

Let Λ be a diagonal matrix $\mathrm{diag}(\lambda(W))$ and V be a matrix whose columns are the eigenvectors in an order corresponding to the order of eigenvalues in $\lambda(W)$. It follows that

$$W = V\Lambda V^{-1}$$

We stabilize W to W_s by scaling all its eigenvalues by its spectral norm if the spectral norm is greater than 1, while maintaining the same eigenvectors, that is,

$$W_s = \begin{cases} V\dfrac{\Lambda}{\rho(W)}V^{-1} = \dfrac{1}{\rho(W)}W & \text{if } \rho(W) > 1 \\ W & \text{otherwise} \end{cases} \qquad (6)$$

Let A_s be the transformed matrix A after stabilization. Plugging in the definition of W, we obtain

$$A_s = \frac{1}{h}\left[\frac{hA + I}{\rho(hA + I)} - I\right]$$

if the spectral norm of W is greater than 1. Replacing A by A_s in Eq. (1), we obtain

$$\mathbf{g}[t+1] - \mathbf{g}[t] = h \left\{ \frac{1}{h} \left[\frac{hA+I}{\rho(hA+I)} - I \right] \mathbf{g}[t] + B \, \mathbf{e}[t] \right\} + \epsilon[t] \qquad (7)$$

There are several theoretical and numerical properties associated with our stabilization strategy. It is evident that any coefficients off the diagonal line in A
with a value close to 0 will be closer to 0 after stabilization. This ensures that no
new interactions between different genes will be introduced by stabilization. The
spectral norm can be found efficiently using the power method without obtaining
the entire eigenvalues or eigenvectors of matrix W. In addition, since there is
no matrix decomposition involved, the stabilized matrix A_s will be real if A is
real, which holds true theoretically but could be violated numerically by other
approaches.

3.4   Statistical Significance of a Discrete Dynamic System Model

Let the minimum p-value of fitting a linear difference equation to gene i be p_i.
The p-value of an entire fitted discrete dynamic system model is computed by

$$\text{p-value} = 1 - \prod_{i=1}^{N}(1 - p_i)$$

where p_i is computed by the F-tests during the fitting of linear model for gene
node i. This defines a conservative p-value since it assumes that the mRNA
levels are independent to each other. Nevertheless, the p-value of a network is
a statistically effective and computationally efficient measure to determine the
chance an estimated system would arise randomly. This p-value is influenced by
1) how well each linear difference equation can be fitted to the data and 2) the
number of nodes in the network, which constitute two competing factors. Our
algorithm minimizes the p-value by trade-off between both factors.

3.5   Implementation and Modeling Details

The network modeling software is written in the R programming language
(R Development Core Team, 2006). For the modeling of the network of 46 genes
shown in Fig. 1, it took about 24 hours on 12 networked computers (Sun Java
Workstation w1100z, Opteron 150 processor, 2.4 GHz clock frequency 1 GB
memory, running 64-bit SuSE Linux [version 10]). The maximum number of
potential regulators including the HMF was set to 5 during the system model
construction.

Acknowledgement. This study was supported in part by NRI Competitive
Grant Program project #ILLR- 2006-02272.

References

YEAst Search for Transcriptional Regulators And Consensus Tracking (YEAS-TRACT), January 2006. URL http://www.yeastract.com. Last Date of Visit: (September 12, 2006)

Akutsu, T., Kuhara, S., Maruyama, O., Miyano, S.: Identification of genetic networks by strategic gene disruptions and gene overexpressions under a Boolean model. Theoretical Computer Science 298(1), 235–251 (2003)

Bonneau, R., Reiss, D.J, Shannon, P., Facciotti, M., Hood, L., Baliga, N.S, Thorsson, V.: The inferelator: an algorithm for learning parsimonious regulatory networks from systems-biology data sets de novo. Genome Biology 7(5), R36 (2006)

Bothast, R., Saha, B.: Ethanol production from agricultural biomass substrate. Adv. App. Microbiol. 44, 261–286 (1997)

Devaux, F., Carvajal, E., Moye-Rowley, S., Jacq, C.: Genome-wide studies on the nuclear PDR3-controlled response to mitochondrial dysfunction in yeast. FEBS Letters 515(1-3), 25–28 (2002)

D'haeseleer, P., Wen, X., Fuhrman, S., Somogyi, R.: Linear modeling of mRNA expression levels during CNS development and injury. In: Pacific Symposium on Biocomputing, pp. 41–52. World Scientific Publishing Co, Singapore (1999)

Edelstein-Keshet, L.: Mathematical Models in Biology. SIAM (2004)

Friedman, N.: Inferring cellular networks using probabilistic graphical models. Science 303, 799–805 (2004)

Golub, G.H., van Loan, C.F.: Matrix Computations, 3rd edn. The Johns Hopkins University Press, Baltimore, MD (1996)

Haugen, A.C., Kelley, R., Collins, J.B., Tucker, C.J., Deng, C., Afshari, C.A., Brown, J.M., Ideker, T., Van Houten, B.: Integrating phenotypic and expression profiles to map arsenic-response networks. Genome Biology 5(12), R95 (2004)

Hegde, P., Qi, R., Abernathy, K., Gay, C., Dharap, S., Gaspard, R., Earle-Hughes, J., Snesrud, E., Lee, N., Quackenbush, J.: A concise guide to cdna microarray analysis. BioTechniques 29, 548–562 (2000)

Imoto, S., Kim, S., Goto, T., Aburatani, S., Tashiro, K., Kuhara, S., Miyano, S.: Bayesian network and nonparametric heteroscedastic regression for nonlinear modeling of genetic network. Journal of Bioinformatics and Computational Biology 1(2), 231–252 (2003)

Lee, J., Godon, C., Lagniel, G., Spector, D., Garin, J., Labarre, J., Toledano, M.B.: Yap1 and Skn7 control two specialized oxidative stress response regulons in yeast. J. Biol Chem. 274(23), 16040–16046 (1999)

Lee, T.I., Rinaldi, N.J., Robert, F., Odom, D.T., Bar-Joseph, Z., Gerber, G.K., Hannett, N.M., Harbison, C.T., Thompson, C.M., Simon, I., Zeitlinger, J., Jennings, E.G., Murray, H.L., Gordon, D.B., Ren, B., Wyrick, J.J., Tagne, J.B., Volkert, T.L., Fraenkel, E., Gifford, D.K., Young, R.A.: Transcriptional regulatory networks in Saccharomyces cerevisiae. Science 298(5594), 763–764 (2002)

Liang, S., Fuhrman, S., Somogyi, R.: REVEAL, a general reverse engineering algorithm for inference of genetic network architectures. Pacific Symposium on Biocomputing 3, 18–29 (1998)

Liu, Z.L.: Genomic adaptation of ethanologenic yeast to biomass conversion inhibitors. Appl. Microbiol. Biotech. 73, 27–36 (2006)

Liu, Z.L., Slininger, P.J.: Development of genetically engineered stress tolerant ethanologenic yeasts using integrated functional genomics for effective biomass conversion to ethanol, CAB International, Wallingford, UK, pp. 283–294 (2005)

Liu, Z.L., Slininger, P.J.: Transcriptome dynamics of ethanologenic yeast in response to 5-hydroxymethylfurfural stress related to biomass conversion to ethanol. In: Recent Research Developments in Multidisciplinary Applied Microbiology: Understanding and Exploiting Microbes and Their Interactions-Biological, Physical, Chemical and Engineering Aspects, pp. 679–684. Wiley-VCH, Chichester (2006a)

Liu, Z.L., Slininger, P.J.: Universal external RNA controls for microbial gene expression analysis using microarray and qRT-PCR. J. Microbiol. Methods, doi:10.1016/j.mimet.2006.10.014 (2006b)

Liu, Z.L., Slininger, P.J., Dien, B.S., Berhow, M.A., Kurtzman, C.P., Gorsich, S.W.: Adaptive response of yeasts to furfural and 5-hydroxymethylfurfural and new chemical evidence for HMF conversion to 2,5-bis-hydroxymethylfuran. J Ind. Microbiol Biotechnol. 31, 345–352 (2004)

Liu, Z.L., Slininger, P.J., Gorsich, S.W.: Enhanced biotransformation of furfural and 5-hydroxy methylfurfural by newly developed ethanologenic yeast strains. Appl Biochem Biotechnol. 121-124, 451–460 (2005)

Lucau-Danila, A., Lelandais, G., Kozovska, Z., Tanty, V., Delaveau, T., Devaux, F., Jacq, C.: Early expression of yeast genes affected by chemical stress. Mol. Cell Biol. 25(5), 1860–1868 (2005)

Luo, C., Brink, D., Blanch, H.: Identification of potential fermentation inhibitors in conversion of hybrid poplar hydrolyzate to ethanol. Biomass Bioenergy 22, 125–138 (2002)

Martin, C., Jonsson, L.: Comparison of the resistance of industrial and laboratory strains of Saccharomyces and Zygosaccharomyces to lignocellulose-derived fermentation inhibitors. Enzy. Micro. Technol. 32, 386–395 (2003)

Meir, E., Munro, E.M., Odell, G.M., von Dassow, G.: Ingeneue: A versatile tool for reconstituting genetic networks, with examples from the segment polarity network. Journal of Experimental Zoology 294, 216–251 (2002)

Ong, I.M., Glasner, J.D., Page, D.: Modelling regulatory pathways in E. coli from time series expression profiles. Bioinformatics 18, S241–S248 (July 2002)

Pal, R., Ivanov, I., Datta, A., Bittner, M.L., Dougherty, E.R.: Generating Boolean networks with a prescribed attractor structure. Bioinformatics 21, 4021–4025 (November 2005)

Palmqvist, E., Almeida, J., Hahn-Hägerdal, B.: Influence of furfural on anaerobic glycolytic kinetics of Saccharomyces cerevisiae in batch culture. Biotechnol Bioeng 62, 447–454 (1999)

R Development Core Team.: R: A Language and Environment for Statistical Computing. R Foundation for Statistical Computing, Vienna, Austria, ISBN 3-900051-07-0, http://www.R-project.org (2006)

Saha, B.: Hemicellulose bioconversion. Journal of Industrial Microbiology and Biotechnology 30, 279–291 (2003)

Schlitt, T., Brazma, A.: Modelling in molecular biology: describing transcription regulatory networks at different scales. Philosophical Transactions of the Royal Society B: Biological Sciences 361(1467), 483–494 (March 2006)

Schmitt, M.E., Brown, T.A., Trumpower, B.L.: A rapid and simple method for preparation of RNA from Saccharomyces cerevisiae. Nucl. Acid Res. 18, 3091–3092 (1990)

Shmulevich, I., Dougherty, E.R., Kim, S., Zhang, W.: Probabilistic Boolean networks: a rule-based uncertainty model for gene regulatory networks. Bioinformatics 18, 261–274 (February 2002)

Taherzadeh, M., Gustafsson, L., Niklasson, C.: Physiological effects of 5-Hydroxymethylfurfural on Saccharomyces cerevisiae. App. Microbiol. Biotechnol. 53, 701–708 (2000)

Takahashi, K.: Multi-algorithm and multi-timescale cell biology simulation. PhD thesis, Keio University, Fujisawa, Japan (2004)

Takahashi, K., Arjunan, S.N.V., Tomita, M.: Space in systems biology of signaling pathways – towards intracellular molecular crowding in silico. FEBS Letters 579, 1783–1788 (2005)

Teixeira, M.C., Monteiro, P., Jain, P., Tenreiro, S., Fernandes, A.R., Mira, N.P., Alenquer, M., Freitas, A.T., Oliveira, A.L., Sá-Correia, I.: The YEASTRACT database: a tool for the analysis of transcription regulatory associations in Saccharomyces cerevisiae. Nucl. Acids Res. 34, D446–451 (2006)

Tomita, M., Hashimoto, K., Takahashi, K., Shimizu, T.S., Matsuzaki, Y., Miyoshi, F., Saito, K., Tanida, S., Yugi, K., Venter, J.C., Hutchison III, C.A.: E-CELL: software environment for whole-cell simulation. Bioinformatics 15(1), 72–84 (1999)

van Kampen, N.: Stochastic Processes in Physics and Chemistry. Elsevier, Amsterdam (1997)

Wahbom, C.F., Hahn-Hägerdal, B.: Furfural, 5-hydroxymethylfurfrual, and acetone act as external electron acceptors during anaerobic fermentation of xylose in recombinant Saccharomyces cerevisiae. Biotechnol Bioeng. 78, 172–178 (2002)

A Computational Approach for the Identification of Site-Specific Protein Glycosylations Through Ion-Trap Mass Spectrometry

Yin Wu[1,5], Yehia Mechref[2,5], Iveta Klouckova[2], Milos V. Novotny[2,5], and Haixu Tang[3,4,5,*]

[1] Department of Computer Science
[2] Department of Chemistry
[3] School of Informatics
[4] Center for Genomics and Bioinformatics
[5] National Center for Glycomics and Glycoproteomics, Indiana University, Bloomington, IN 47408
Tel.: 812-856-1859
hatang@indiana.edu

Abstract. Glycosylation is one of the most common post-translational modifications (PTMs) of proteins, the characterization of which is commonly achieved utilizing mass spectrometry (MS). However, its applicability is currently limited by the lack of computational tools capable of autmoated interpretation of high throughput MS experiments which would allow the characterization of glycosylation sites and their microheterogeneities. We present here a computational approach which overcomes this problem and allows the identification and assignment of the microheterogeneities of glycosylation sites of glycoproteins from liquid chromatography ion-trap-based mass spectrometry (LC/MS) data. This method was implemented in a software tool and tested on several model glycoproteins. The results demonstrate the potential of our computational approach in automating the high throughput identification of glycoproteins.

Keywords: Glycoproteomics; Mass spectrometry; Site-specific Glycosylation; Algorithm.

1 Introduction

With approximately 50% of all proteins now considered to be glycosylated [1], this type of PTM is widespreaded and physiologically important in mammalian systems involved in many functions such as cell-cell recognition and protein-protein intercations which mediate many physiological functions. A growing list of glycoproteins has been recognized to act through a recognition of oligosaccharide

---

* Corresponding author.

T. Ideker and V. Bafna (Eds.): Syst. Biol. and Comput. Proteomics Ws, LNBI 4532, pp. 96–107, 2007.
© Springer-Verlag Berlin Heidelberg 2007

chains and their microheterogeneities at the site of modification. Consequently, aberrant glycosylation has now been recognized as an attribute of many mammalian diseases, including hereditary disorders, immune deficiencies, neurodegenerative diseases, cardiovascular conditions, and cancer [2, 3]. Glycosylation of a protein is accomplished through linkage to Asn residues (designated as N-glycosylation) or to Ser/Thr residues (designated as O-glycosylation). Unlike nucleic acids and proteins, glycans are synthesized in a template-free fashion by a number of glycotransferases. All N-linked glycans share a common core structure, called the "pentamer", consisting of two N-acetylglucoseamin (GlcNAc) residues and three mannose residues. Additional monosaccharides can be further linked to this core structure to form diverse branching glycan structures. On the other hand, O-linked glycans lack a common core structure, and have higher sequence diversity, yet they are commonly short. In addition to the lack of synthesis template, the characterization of glycosylation is further complicated by branching and anomericity as well as the presence of several glycan structures attached to the same glycosylation sites which is commonly referred to as "microheterogeneity".

In recent years, mass spectrometry (MS) has been used to determine both the sequence/structure analysis of glycans in the glycoproteins [4, 5], and site-specific analysis of glycoproteins [6, 7]. As a commonly used MS platform, trypsin digestion followed by liquid chromatography mass spectrometry (LC/MS) has been used to analyze glycoproteins at high throughput and high sensitivity [7, 8]. Numerous attempts have been made [9] to develop enrichment methods for glycoproteins from complex biological samples. Logically, the great majority of these enrichment methodologies rely on the use of immobilized lectins, which in their modern versions permit a more or less selective enrichment of the pools of glycoproteins for proteomic/glycomic studies [10-14]. However, the applicability of these approaches is still restricted, largely due to the lack of a computational method allowing automatic identification of glycopeptide signals in a complex sample. Therefore, the characterization of the glycosylation site of a protein remains one of the great challenges in MS-based proteomics [15].

Although a lot of efforts has been focused on the development of analytical tools which permits automated analysis of glycans at high throughput using mass spectrometry (see [16] for a review), there has been so far no software tool developed for assigning protein glycosylation sites from high throughput proteomics experiments. In this paper, we present a computational approach to the glycoprotein identification, focusing on the data generated by a recently developed glycoprotein analysis protocol based on low mass-resolution LC/MS and collision-induced dissociation tandem MS (CID-MS/MS) of instrument-selected glycopeptides [8]. Our approach integrates two scoring schemes in glycopeptide identification, which capture the co-eluted glycopeptide ions in MS spectra and fragmentation patterns of glycopeptides in MS/MS spectra, respectively. We implemented this approach in a computational tool and tested it on several model glycoproteins. We show that our program can identify the glycopeptides, the potential glycosylation sites as well as the microheterogeneities of each site.

2 Methods

2.1 Sample Preparation and Analysis

We obtained three mass spectrum data sets from three glycoprotein samples which were tryptically digested. A 10-µL aliquot of a glycoprotein solution (0.1 µg/µL in 50 mM ammonium bicarbonate) was thermally denatured at 95°C for 15 min, centrifuged and cooled down to room temperature. The sample was reduced with dithiothreitol, at 50°C, and alkylated with iodoacetamide, in the dark, at room temperature. The reaction mixture was incubated after the addition of trypsin stock solution at a 50:1 ratio (protein:trypsin). To generate the deglycosylated sample, the trypitcally digested samples were incubated with PNGase F at 37°C overnight. A nanoLC pulled-tip column was utilized and directly coupled to LCQ Deca XP ion-trap mass spectrometer (ThermoElectron, San Jose, CA), which was operated in the positive-ion mode. The instrument recorded full mass spectra (m/z 250-2000), MS/MS data-dependent spectra and mass spectra of activated ion source CID at different electrical potentials. The Q activation values for all MS^n were set to 0.250, while the activation times were set to 75 msec.

2.2 Computational Methods

We first evaluate the probability of each obtained MS/MS spectrum to be that of a glycopeptide. The CID spectrum of a glycopeptide often features a series of strong ion peaks corresponding to a partial or complete loss of the glycan side chain due to the fragmentation at a glycosidic bond [18]. These peaks usually are much stronger than the peaks resulted from peptide backbone cleavages, hence, can be easily distinguished from a glycopeptide spectrum (e.g. see **Figure 1A**). Even though it is hard to reconstruct the complete glycan structure from these fragment ions, one can reconstruct a *sequence tag* of oligosaccharide, i.e. a series of strong peaks corresponding to consecutive monosaccharide cleavages. Similar to the sequence tag approach commonly used in peptide identification [19-21], we have utilized a dynamic programming algorithm to find the strongest oligosaccharide sequence tag from a given MS/MS spectrum.

We build a spectrum graph [22], in which each node represents a peak in the spectrum and an edge is linked between two nodes, if the mass difference between the two corresponding peaks is equal[1] to the mass of a particular monosaccharide (**Figure 1A**). Unlike the spectrum graph used for peptide *de novo* sequencing, we do not create nodes to represent mass 0 and the parent mass. To weight the spectrum graph, we tried two schemes. The first gave each edge the same weight, while the second weighted each edge by the average intensity of the two peaks that are linked. We then find the longest path in the spectrum graph using a dynamic programming algorithm [23]. We note the longest path can start at any node and end at any node in the spectrum graph, thus representing a sequence tag (of oligosaccharide), not a complete construction of the entire glycan structure.

---

[1] Throughout this paper, we say two masses are equal if their difference is within the resolution of a MS instrument (0.7 for ion trap MS used in this paper).

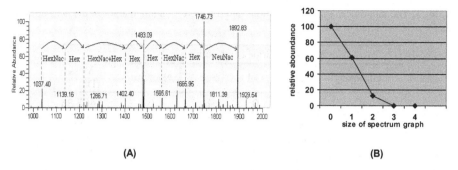

Fig. 1. Evaluating the probability of an MS/MS spectrum to be originating from a glycopeptide. (A) The strongest oligosaccharide sequence tag was identified using the dynamic programming algorithm to find the longest path in the spectrum graph; (B) The probability distribution of the length of oligosaccharide sequence tag obtained from the statistics on 1320 MS/MS spectra of deglycosylated peptides.

Next, we evaluate the probability of an MS/MS spectrum being a glycopeptide spectrum by the length of the oligosaccharide tags we generate. **Figure 1B** shows the probability distribution in the length of sequence tags that can be found from MS/MS spectra of deglycosylated peptides, which is obtained through the statistics on a reference LC/MS data set, generated for the deglycosylated forms of model proteins (see section Data acquisition for experimental details). It is clear that the random probability of finding an oligosaccharide sequence tag with length 4, i.e. four peaks corresponding to three consecutive monosaccharide cleavages, is very low (0.17%), whereas such a tag can be found in most (75%) glycopeptide spectra. The random probabilities from this statistics were used to assign a P-value for the evaluation of each MS/MS spectrum.

For each glycosylation site in a glycoprotein, there are often several structural glycan variants, which are referred to as *site-specific microheterogeneities* [6]. As a result, the mixture, resulting from a proteolytic digestion of a glycoprotein, often consists of different *clusters of peptide glycoforms*, i.e. glycopeptides with the same peptide backbone but different glycans; this has often important biological consequences. In addition to their complexity, peptide glycoforms provide valuable information for helping to identify glycopeptides, because a cluster of peptide glycoforms usually co-elutes in an LC/MS run owning to their common chemical properties [24].

Suppose M is a set of ion masses within an elution time window, $M = \{s_1, s_2, ...,s_{|M|}\}$. If a subset of M corresponds to the masses of a cluster of glycoforms, it is likely that this subset represents microheterogeneities of the same glycosylation site. Let G be the set of the masses of the possible glycans attached to the same peptide backbone, $G = \{m_1, m_2, ...,m_{|G|}\}$. G can be deduced from the monosaccharide composition for N-glycans, which is studied in this paper (**Table 1**). For O-glycans, one needs to perform a separate O-glycan profiling procedure to obtain potential glycoforms in a sample of interest [25].

Table 1. The monosaccharide compositions and the corresponding masses of the N-glycoforms that were considered in this study[2]

Monosaccharide composition					Mass (Da)
GlcNAc	Man	Gal	Fuc	NeuNAc	
4	3	0	1	0	1463.36
4	3	1	1	0	1625.50
4	3	2	0	0	1641.50
4	3	2	1	0	1787.65
4	3	2	0	1	1932.76
4	3	2	1	1	2078.90
4	3	2	0	2	2224.02
5	3	3	0	1	2298.10
4	3	2	1	2	2370.16
5	3	3	1	1	2444.24
5	3	2	1	2	2573.36
5	3	3	0	2	2589.36
5	3	3	1	2	2735.50
5	3	3	0	3	2880.61
5	3	3	1	3	3026.76
5	3	3	0	4	3171.87
6	3	4	0	3	3245.95
6	3	4	1	3	3392.09
5	3	3	0	5	3463.13
6	3	4	0	4	3537.21
6	3	4	1	4	3683.35
6	3	4	2	4	3829.50

We intend to select a subset of M, denoted as $M_c \subset M$, such that it can be represented as $\{M_p+m_1, M_p+m_2, \ldots, M_p+m_k\}$, in which $k \leq |G|$, is the number of identified peptide glycoforms; M_p is the (unknown) mass of peptide backbone; and $m_i \in G$, for $i = 1, 2, \ldots, k$. We note that the only variable here that governs the selection of subset M_c is the peptide backbone mass M_p; once M_p is determined, we then select the largest subset of M for M_c, which match the maximal number of glycan masses in G. Therefore, to evaluate all potential subsets M_c of M satisfying the above condition, we just need to evaluate each possible M_p. We compute a convoluted spectrum for the whole potential range of M_p, $S(M_p)$, integrating the clustering property of glycoforms, and the P-values of MS/MS spectra (if any) associated with the peaks in M, by a spectrum convolution algorithm, which is depicted in **Figure 2**. [22, 26]:

---

[2] In the actual implementation, three distinct charge states (+2, +3 and +4) were assigned to each glycoform and the resulting mass/charge ratios were used to the glycoform mass set for spectrum convolution.

$$S(M_p) = \sum_{i=1}^{|G|} \sum_{j=1}^{|M|} \left(P_1(s_i) \cdot P_2(m_j) \cdot l(s_i - m_j - M_p) \right)$$ (1)

where,

$$l(x) = \begin{cases} 1 & if\ x = 0 \\ 0 & otherwise \end{cases}$$ (2)

and $P_1(s)$ is the scoring for ion peak s: $P_1(s) = P_1^A(s) \cdot P_1^B(s)$, where $P_1^A(s)$ is equal to the P-value that is assigned to the MS/MS spectrum(if the MS/MS spectrum does not exists for s, it is assigned to a prior probability) and $P_1^B(s)$ is a score based on the intensity of the ion peak s. $P_2(m)$ is the scoring for a glycan structure, which is assigned to be a constant in this study, but in general can be assigned by the expectation of observing this glycan in the sample based on either a prior probability estimation or a prior knowledge of the possible glycan structures associated with the glycoprotein being characterized.

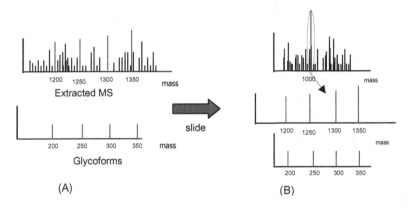

Fig. 2. Spectrum convolution. (A) The set of glycoform masses is slided (convoluted) over an extracted MS spectrum within a certain elution time window (+/-1.5 mins). A convolution score $S(M_p)$ is computed for each sliding offset M_p. (B) A high peak in the resulting convoluted spectrum (uppermost spectrum) correspond to a good match between the MS spectrum and glycoform mass set (two lower spectra).

$P_1^A(s)$ is computed based on length distribution of the length of the monosaccharide path in a deglycosylated reference data set. We assume that it follows a normal distribution, and the mean and variance of the normal distribution is calculated. The prior probability is assigned as $P_1^A(s_\mu)$, where s_μ is the mean value of the path lengths of the spectrums in the deglycosylated data set.

$P_1^B(s)$, of an MS/MS spectrum, is computed based on the intensity distribution of the noise associated with the analysis. This is accomplished through evaluating the MS spectra in the first 5 minutes and the last 5 minutes of the LC run, where commonly no peptide is present. These spectra are overlapped into one and assigned to be the background noise. Again, we assume that the intensity of noise peaks follows a normal distribution, and the mean and variance of this normal distribution is calculated.

In a real LC/MS experiment, one glycopeptide may carry different electric charges and, hence, appear as different ions in the spectrum. Therefore, we consider various charge states (+2, +3 and +4) in the glycoform mass set. Three mass/charge ratios are then assigned to each glycoform and all included in the mass set for the convolution analysis. This is needed since this study is aimed at utilizing a data generated from a low mass-resolution instrument; however, this might not be needed in the case of high mass-resolution instrument.

We generalized the spectrum convolution algorithm described above to a scanning of the full range of elution time in an LC/MS experiment. We used a sliding window with the size of 3.0 (+/-1.5) minutes to scan the whole elution time range. All ions (and their associated MS/MS spectra) were collected within the time window. The sequence-tag-finding algorithm was applied to each collected MS/MS spectrum, and then the spectrum convolution algorithm was applied to all collected peaks. Each potential peptide mass M_p was given a score $S(M_p, e)$ at each elution time e.

$S(M_p, e)$ can be represented as a two dimensional map (**Figure 3A**), in which a dark point represents a potential cluster of peptide glycoforms with corresponding peptide mass M_p eluting together at the time e, with the grayscale representing the score $S(M_p, e)$ received from the scanning. It can be seen from the map that a cluster of peptide glycoforms may elute for a short period of time (up to 5 minutes), which is then shown as a short vertical line in the two-dimensional map (**Figure 3A**). Note that even though this two-dimensional map looks similar to the real time LC/MS map, it is a virtual map. However, one can compare this map with a real-time map from deglycosylated form of the sample [27] to identify the peptide that is site-specifically glycosylated.

3 Results and Discussion

We present a computational approach to glycoprotein identification and we report an implementation of this approach for the analysis of different glycoproteins. We implemented it into a computational tool GlyPID in C++ with QT library. To exploit its potential in glycoprotein identification, we initially test it on three model glycoproteins: human immunoglobulin G (IgG), bovine fetuin and human α_1-acid glycoprotein (AGP). In this study, we focus on the N-glycosylation sites of these proteins, because N-glycans have relatively regular monosaccharide compositions. The method can be extended to the analysis of O-glycosylation sites, if the glycoforms in the sample are deduced from a separate O-glycan profiling analysis.

Figure 3A shows the interpretation results of GlyPID for bovine fetuin, which has three annotated (N_{156}, N_{99}, N_{176}) N-glycosylation sites, while **Figures 3B** depicts the tandem MS spectra of representative of one of the three glycosylation sites associated with fetuin.

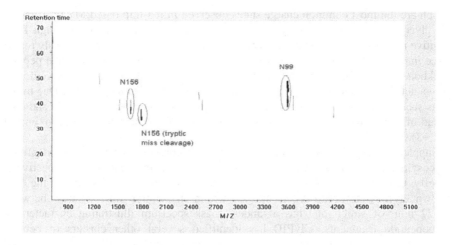

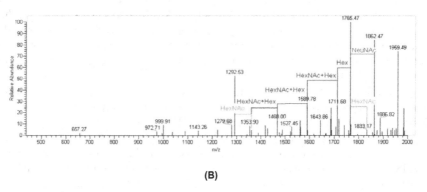

(B)

Fig. 3. Putative glycopeptides predicted by GlyPID. (A) Two dimensional map from the full elution time scanning for model glycoprotein Bovine fetuin. The Y-axis represents the elution time. The X-axis represents the mass of the peptide backbone. The peptide backbones corresponding to three known glycopeptides were highlighted in the map. They received three highest scores in the whole scan. (B) MS/MS spectrum associated to one of the three glycopeptides. The peaks of the identified monosaccharide multi-tag are annotated in color.

The three top scored clusters (each with at least 8 peptide glycoforms) are shown in supplementary **Table I**. The data illustrates the clusters from two of the three known glycosylation sites which received the top scores in the whole time scanning, which provides the proof-of-concept that our computational method can automatically identify glycopeptides from high-throughput LC/MS experimental data. Besides that, a peptide containing one miss-cleavage of site N_{156} also received high score. The sequence of this partially-cleaved peptide is KLCPDCPLLAPLN$_{156}$DSR. Glycosylation site (N176) was not assigned, since there is no high quality MS/MS spectrum with lower charge states acquired by MS instrument for any glycopeptide covering this site. In fact, there is a high quality MS/MS spectrum with charge +5 that was later confirmed by manual inspection to be that of a tryptic glycopeptide covering this site. However, currently GlyPID only considers charge states +2, +3 and +4, which are the most common charge states observed in ion-trap based MS instruments for glycopeptides. Incorporating higher charge state, e.g. +5, will create high false positive rates (data not shown). We emphasize that GlyPID can simultaneously report three important features of glycoproteins: the glycosylation site (through the peptide backbone mass), the microheterogeniety of the glycosylation (i.e. the glycoforms associated to the glycosylation site) and the elution time of glycopeptides. Due to the facts such as instrumental error (-0.7, +0.7), maximum ion charge level (+4) and data round off, the error window of predicted mass of peptide backbone is approximated (-3, +3).

Supplementary **Tables II** and **III** give a complete report of the identified glycosylation sites and their microheterogeneties for IgG and AGP respectively. Briefly, GlyPID identified a cluster of peptide glycoforms corresponding to the only known glycosylation site (N_{297}) in IgG, within the elution time window between 18-22 min, of which one has a tandem mass spectrum illustrating characteristic glycopeptide fragments. GlyPID has identified several other clusters of peptide glycoforms, of which several received even higher scores than that of the annotated site and associated to at least one well scored MS/MS spectrum. Some of them correspond to the partially cleaved tryptic glycopeptides covering the same site. For example, one of the identified clusters is associated to a partially cleaved tryptic peptide TKPREEQYN$_{297}$STYR, which is found within the elution time window between 18.5-22 min.

AGP has five annotated N-glycosylation sites (N_{33}, N_{56}, N_{72}, N_{93}, N_{103}). GlyPID can identify 3 distinct clusters of glycoforms within the elution time between 16.5-18.5, 32.5-35.5 and 40.5-43.5 min, which corresponds to four glycosylateion sites (N_{56}, N_{72}, N_{93}, N_{103}), respectively, among several other clusters. Each of these clusters consists of 10-14 peptide glycoforms, while at least one of them has a identified glycopeptide MS/MS spectrum. Two clusters of peptide glycoforms (corresponding to N_{72} and N_{93}) happen to have similar elution time as well as peptide backbone masses, thus indistinguishable in our analysis. The fifth glycosylation site (N_{33}) is not identified by our method due to the low abundance of the peptide glycoforms. In addition, GlyPID also identified a cluster of glycoforms corresponding to a partially cleaved peptide SVQEIQATFFYFTPN$_{72}$KTEDTIFLR covering glycosylation site N_{72} of AGP.

The preliminary evaluation of our method on several model glycoproteins has demonstrated its high sensitivity and applicability to high-throughput glycoproteomics projects. We are now applying this computational tool to analyze complex proteome samples, e.g. human blood serum proteome, in an attempt to identifying glycoprotein biomarkers and their glycosylation sites and microheterogeneneity.

Although we focus on the N-glycosylation in this study, it should be fairly straightforward to extend the utility of this algorithm to O-glycosylations. The sensitivity, however, may be reduced, owning to the much higher sequence diversity for O-glycans. One possible way to address this issue is to conduct a separate O-glycan profiling, allowing the identification of glycan structures associated with the glycoproteins. This information could be utilized then to derive the potential peptide glycoforms originating from O-glycans. Currently, GlyPID has an implemented function that allows the user to define the optional masses of the glycan structures associated with the glycoproteins analyzed. Ultimately, we want to integrate this tool with other tools that we developed for glycan structure analysis [17] towards a toolbox for automated glycomic and glycoproteomic approaches.

After we identify the glycopeptides (i.e. determine the mass of peptide backbone), it is also important to determine the sequence of the peptide. One way of achieving this is to use the MS^3 spectra that are generated from the fragmentation of ions representing the peptide backbone with one GlcNAc residue attached observed as the most intense ion in the middle of the MS/MS spectra of glycopeptide. A regular database searching procedure of this MS^3 can be used to identify these peptides [8]. This approach is completely dependent on the sensitivity of mass spectrometers and their ability to generate reliable MS^3 data. Another approach to identifying the peptides is to use a reference data set generated for the deglycosylated forms of sample. Since it is observed that the deglycosylated form of a glycopeptide often elute at a similar time as the glycopeptide, we can potentially identify the peptide using the MS/MS spectra in the reference set, with a predicted backbone mass M_p around the same elution time. We are exploring these approaches and will intend to include them in future software.

Acknowledgment

We thank Vijetha Vemulapalli for providing us the visualization toolkit for the two dimensional map. This work was supported by Grant No.GM24349 from the National Institute of General Medical Sciences, U.S. Department of Health and Human Services and a center grant from the Indiana 21st Century Research and Technology Fund. The mass spectrometer used in this study was acquired as a result of support from Indiana Genomics Initiative (INGEN), which is funded in part by the Lilly Endowment, Inc. This work was also supported by the National Center for Glycomics and Glycoproteomics funded by NIH-National Center for Research Resources (NCRR).

Supplementary Materials

Please check the following URL for the supplementary materials.
http://darwin.informatics.indiana.edu/applications/glypid/LNBI_2006/
supplementary_materials.htm

Complementary Website

http://darwin.informatics.indiana.edu/applications/glycomics/

References

1. van den Steen, P., et al.: Crit. Rev. Biochem. Mol. Biol. 33, 151–208 (1998)
2. Dennis, J.W., Granovsky, M., Warren, C.E.: Protein glycosylation in development and disease. Bioassays 21, 412–421 (1999)
3. Lowe, J.B., Marth, J.D.: A genetic approach to mammalian glycan function. Ann. Rev. Biochem. 72, 643–691 (2003)

4. Mechref, Y., Novotny, N.V., Krishnan, C.: Structural characterization of oligosaccharides using MALDI-TOF/TOF tandem mass spectrometry. Anal Chem. 75(18), 4895–4903 (2003)

5. Zaia, J.: Mass spectrometry of oligosaccharides. Mass Spectrom Rev. 23(3), 161–227 (2004)

6. Novotny, M.V., Mechref, Y.: New hyphenated methodologies in high-sensitivity glycoprotein analysis. J. Sep. Sci. 28(15), 1956–1968 (2005)

7. Wuhrer, W., Deelder, A.M., Hokke, C.H.: Protein glycosylation analysis by liquid chromatography-mass spectrometry. J. Chromatogr B. Analyt Technol Biomed Life Sci. 825(2), 124–133 (2005)

8. Mechref, Y., Klouckova, I., Novotny, M.V.: Ion-trap-based strategy for the mass-spectrometric assignment of glycosylation sites in proteins. Rapid Commun. Mass Spectrom (submitted) (2006)

9. Hjorth, R., Vretblad.: Group fractionation of human serum glycoproteins using Sepharose bound lectins. In: Lect. Chem. Soc. Int. Symp. Uppsala, Sweden, Ellis Horwood Ltd (1976)

10. Hage, D.S.: Affinity Chromatography: A Review of Clinical Applications. Clin. Chem. 45(5), 593–615 (1999)

11. Nawarak, J., Phutrakul, S., Chen, S.-T.: Analysis of Lectin-Bound Glycoproteins in Snake Venom from the Elapidae and Viperidae Families. J. Proteome Res. 3(3), 383–392 (2004)

12. Xiong, L., Andrews, D., Regnier, F.: Comparative Proteomics of Glycoproteins Based on Lectin Selection and Isotope Coding. J. Proteome Res. 2(6), 618–625 (2003)

13. Yang, Z., et al.: A study of glycoproteins in human serum and plasma reference standards (HUPO) using multilectin affinity chromatography coupled with RPLC-MS/MS. Proteomics 5, 3353–3366 (2005)

14. Madera, M., Mechref, Y., Novotny, M.V.: Combining Lectin Microcolumns with High-Resolution Separation Techniques for Enrichment of Glycoproteins and Glycopeptides. Anal. Chem. 77(13), 4081–4090 (2005)

15. Reinders, J., et al.: Challenges in mass spectrometry-based proteomics. Proteomics, 4(12), 3686–3703 (2004)

16. Lieth, C.W.v.d., Lutteke, T., Frank, M.: The role of informatics in glycobiology research with special emphasis on automatic interpretation of MS spectra. Biochim Biophys Acta. Epub in advance (2005)

17. Tang, H., Mechref, Y., Novotny, M.V.: Automated interpretation of MS/MS spectra of oligosaccharides. Bioinformatics 21(1), i431–i439 (2005)

18. Mechref, Y., Novotny, M.V.: Structural investigations of glycoconjugates at high sensitivity. Chem Rev. 102(2), 321–369 (2002)

19. Mann, M., Wilm, M.: Error-tolerant identification of peptides in sequence databases by peptide sequence tags. Anal Chem 66(24), 4390–4399 (1994)

20. Sunyaev, S., et al.: MultiTag: multiple error-tolerant sequence tag search for the sequence-similarity identification of proteins by mass spectrometry. Anal Chem 75(6), 1307–1315 (2003)

21. Frank, A., et al.: Peptide sequence tags for fast database search in mass-spectrometry. J. Proteome Res. 4(4), 1287–1295 (2005)

22. Dancik, V., et al.: De novo peptide sequencing via tandem mass spectrometry. J Comput Biol. (1999)

23. Skiena, S.: The Algorithm Design Manual (1998)

24. Huang, Y., Mechref, Y., Novotny, M.V.: Microscale Nonreductive Release of O-linked Glycans for Subsequent Analysis Through MALDI Mass Spectrometry and Capillary Electrophoresis. Anal. Chem. 73, 6063–6069 (2001)
25. Peter-Katalinic, J.: O-glycosylation of proteins. Methods Enzymol. 405, 139–171 (2005)
26. Pevzner, P., et al.: Efficiency of database search for identification of mutated and modified proteins via mass spectrometry. Genome Res. (February 2001)
27. Leptos, K.C., et al.: MapQuant: Open-source software for large-scale protein quantification. Proteomics, Epub in advance (2006)

De Novo Signaling Pathway Predictions Based on Protein-Protein Interaction, Targeted Therapy and Protein Microarray Analysis

Derek Ruths[1,*], Jen-Te Tseng[2], Luay Nakhleh[1], and Prahlad T. Ram[2]

[1] Department of Computer Science, Rice University, Houston, TX 77005, USA
[2] University of Texas M.D. Anderson Cancer Center, Houston, TX 77030, USA
druths@cs.rice.edu

Abstract. Mapping intra-cellular signaling networks is a critical step in developing an understanding of and treatments for many devastating diseases. The predominant ways of discovering pathways in these networks are knockout and pharmacological inhibition experiments. However, experimental evidence for new pathways can be difficult to explain within existing maps of signaling networks.

In this paper, we present a novel computational method that integrates pharmacological intervention experiments with protein interaction data in order to predict new signaling pathways that explain unexpected experimental results. Biologists can use these hypotheses to design experiments to further elucidate underlying signaling mechanisms or to directly augment an existing signaling network model.

When applied to experimental results from human breast cancer cells targeting the epidermal growth factor receptor (EGFR) network, our method proposes several new, biologically-viable pathways that explain the evidence for a new signaling pathway. These results demonstrate that the method has potential for aiding biologists in generating hypothetical pathways to explain experimental findings.

Our method is implemented as part of the PathwayOracle toolkit and is available from the authors upon request.

1 Introduction

Altered cellular signaling networks can give rise to the oncogenic properties of cancer cells [8], increase a person's susceptibility to heart disease [6], and are responsible for many other devastating diseases [8,3]. As a result, major efforts are currently underway to establish high-resolution maps of signaling networks for various disease-causing cells. These can be used to inform the development of diagnostic methods and pharmacological treatments.

In the laboratory, targeted manipulation experiments either using knockouts (i.e., siRNA or genetic knockout organisms) or pharmacological agents are a primary method for uncovering new connectivity or parts of a signaling network. The goal of such experiments is to generate results that cannot be explained using existing signaling pathway maps or models. These results are important because they signal the discovery of new pathways, but at the same time raise the very open-ended issue of identifying the cause of the incongruous result.

---

* Corresponding author.

T. Ideker and V. Bafna (Eds.): Syst. Biol. and Comput. Proteomics Ws, LNBI 4532, pp. 108–118, 2007.
© Springer-Verlag Berlin Heidelberg 2007

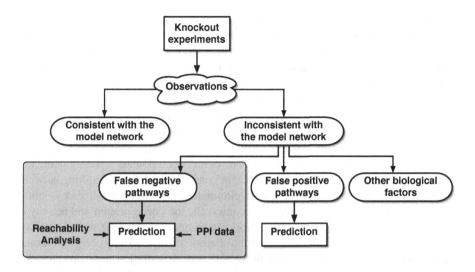

Fig. 1. The path from experiment to new biological insights. Informative knockout or inhibition results are those that cannot be explained by the model. Once such a result has been obtained, the biologist must consider the possible causes for the inconsistency. This paper handles the case of an incomplete signaling model (in the grey box) by providing a computational method for detecting absent pathways and predicting new ones.

As shown in Fig. 1, several explanations can account for unexpected results:

1. *The model is missing signaling pathways.* In this situation, the result is unexpected because interaction paths exist in the biological signaling network that are not represented in the model. These missing paths are false negatives since the model indicates that no such paths exist.
2. *The model contains incorrect signaling interactions or pathways.* Particularly when dealing with diseased cells, signaling network models based on different cell lines can be inaccurate: interactions in one cell line may not exist in the diseased network under study. Thus, the model contains paths that are false positives—paths that do not exist in the context of the cell being studied.
3. *Biological factors have influenced the result.* These can range from technical challenges such as experimental conditions to issues of great scientific importance such as a lack of specificity in the drug being used to knockout or inhibit part of the network.

Thus, when faced with an unexpected result from a knockout or inhibition experiment, the biologist has a large space of potential causes that he or she must consider. As a result, there is a significant need to develop tools that expedite the process of generating hypotheses to explain unexpected targeted manipulation experimental results.

In this paper, we present a novel computational method for identifying and handling knockout or inhibition results that belong to the first class discussed above—those that cannot be explained because the model is missing pathways. Our method (1) identifies results for which the model network is missing paths and (2) generates biologically-viable pathways that can explain the result. These generated pathways become hypotheses that the biologist can then use as a basis for further experiments or as paths that are added to the existing network model. Prior work in this area has focused on related problems in the transcriptional network domain [20,21]. However, to our knowledge, this method is the first to use knockout or inhibition experiments to guide the prediction of missing pathways in the cellular signaling network.

In order to generate new pathways, our approach integrates knockout or inhibition result data with protein interaction data—both sources of information about interactions that occur in signaling networks.

In a knockout or inhibition experiment, one or more compounds in the signaling network are rendered inactive through chemical or genetic means. In the resulting network, any role that these compounds played are eliminated. The modified network is stimulated and set into motion. At various time intervals, the concentration and activity of various proteins within the modified network are compared to those in the original network. A statistically significant change in the concentration or activity of a given protein in the modified network indicates that this protein and the inhibition target must interact. A reasonable representation of such a positive result is the knowledge that a protein X interacts with another protein Y. Since this captures the interaction information supplied by the experiment, this is the representation we use throughout this paper.

Protein interaction data, commonly stored in protein-protein interaction databases, is another major source of interaction information. This data is primarily generated by high-throughput experimental methods that identify protein pairs that are likely to interact. Unlike the results of knockout or inhibition experiments, all interactions returned by these high-throughput methods are putative. As a result, the false positive rate in protein interaction databases has been shown to be high [15]. Various methods, ranging from literature search to comparisons across organisms, have been proposed for assessing the likelihood of an interaction being correct [9,4,2,18,16]. When a protein interaction database is coupled with an interaction confidence measure, it becomes a useful source of information on interactions that occur within the cell.

Since signaling networks ultimately are massive webs of directed protein interactions, one might expect that new signaling topology could be uncovered by dissecting these protein interaction databases. Yeang et al. considered this question with respect to transcriptional networks [20]. In a more recent study, Scott et al. [15] considered this problem with respect to signaling networks and found that highly biologically-relevant topologies could be extracted from these interaction networks. In their analysis, they recovered the MAP kinase and ubiquitin-ligation signaling pathways from a computational search of the MIPS interaction database [12].

Our approach uses this idea of discovering topological structure within a protein interaction dataset by considering it within the context of a single knockout or inhibition experiment. The computational technique searches a protein interaction network for biologically-viable pathways that account for the results of the experiment. We make the assumption that interactions with a high likelihood of being correct are biologically-viable. Extending this assumption to the pathway-level, we consider a pathway to be biologically-viable if the product of the likelihoods of each interaction in the pathway is high. Therefore, our method searches a protein interaction network for the best supported interaction paths that connect X and Y.

In order to test our method, we experimentally and computationally determined the effect of pharmacological inhibitors on changes in signaling network function in human breast cancer cells. Two human breast cancer cell lines were treated with three different pharmacological inhibitors targeting different signaling molecules. We found an unexpected inhibitory interaction between *MEK1* and *c-Src*. Given this result, our method generates excellent candidate pathways that explain the observed knockout or inhibition pattern and are consistent with other biologically known properties of the *EGFR* network. This result can be taken as evidence that our method's generated pathways can be considered reasonable hypotheses for the true signaling network topology underlying experimental results.

In order to make our method available for use, we have implemented it as a Java tool and bundled it with the PathwayOracle software package. PathwayOracle is available upon request from the authors.

2   Results and Discussion

2.1   Experimental Results

In order to understand how targeted manipulations alter different nodes in the signaling network we used inhibitors to specific molecules and measured changes in several proteins within the network using protein microarrays. Combining targeted pharmacological manipulations with protein array technology allows us to simultaneously measure changes in a large number of signaling molecules very rapidly. Using this method we treated breast cancer calls with three inhibitors of the signaling network.

The inhibitors used were Iressa (EGFR kinase inhibitor), perifosine (AKT inhibitor) and PD98059 (MEK inhibitor). Iressa is currently used in clinical treatment of patients, and AKT and MEK inhibitors are in pre-clinical and early phase clinical trials [7].

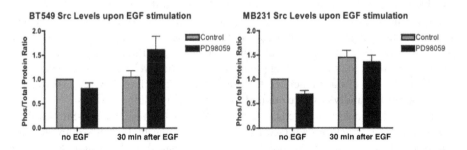

Fig. 2. Experimental microarray data from BT549 and MDA-MB-231 breast tumor cells treated with the MEK1 inhibitor PD98059 shows that the level of phopho c-Src is increased in BT549 cells but not in MDA-231 cells upon EGF stimulation. The two graphs show the phospho c-Src levels in the two cell lines after normalization for protein loading, the first bar corresponds to control cells and the second bar corresponds to cells treated with the MEK1 inhibitor for 30 minutes.

Analysis of the data from the two cell lines at two different time points in which post stimulation revealed changes in signaling within the network (see Figure 2). We observed the expected changes (not shown), i.e. when the MEK inhibitor was used EGF did not stimulate MAPK1,2 but the activation of AKT was not altered. When Iressa was used to inhibit EGFR the activation of MAPK 1,2, was blocked in response to EGF in Ras wild type cells but not in cells with a Ras activation mutation. Similarly Iressa blocked AKT activation of PTEN wild type cells but not in PTEN deletion cells. Having observed expected outcomes we were very intrigued by results that were unexpected. For example we found that in BT549 breast tumor cells PD 98059 elevated c-Src basal phosphorylation levels in EGF stimulated cells. However, this was not the case in MDA-MB-231 cells, where there was no increase in c-Src phosphorylation when compared to control. This data suggests that by inhibiting MEK1 we are also increasing c-Src. There could be two explanations for this result, the first being that MEK and c-Src are connected through a signaling pathway in BT549 cells, or the second being that the MEK inhibitor has non-specific activity on c-Src. However, based on the result in MDA-231 cells where there is no increase in c-Src it does not appear that there is a non-specific drug effect on c-Src. From these results we checked our existing signaling network model to find connectivity between MEK1 and c-Src, and found no existing pathway.

2.2 Pathway Prediction Results

From our experimental data we observe that inhibiting MEK1 results in an increase in phosphorylation of c-Src in BT549 cells. In order to understand how inhibiting MEK1 could activate c-Src we performed a PubMed search and found no previously published work describing MEK1 activation of c-Src. There were several publications showing that c-Src could activate MEK1, but not vice versa.

Ordinarily when faced with this scenario of having an unexplained experimental outcome and no previously described pathway from MEK1 to c-Src, the biological investigator is faced with hours of literature searches in an attempt to find pair-wise interactions that can connect MEK1 to c-Src. These searches frequently result in several possible best guess pathways that the investigator would then have to check individually. This method of going down a laundry list of pathways to test is very inefficient and uses valuable time, manpower and resources. Computational methods to identify possible pathways focus this effort and allow the investigator to logically rank and test the pathways based on the modeling prediction. We have developed such a method and show here the use of our model and the use of iterative cycling between experiments and modeling to rapidly advance our understanding of signaling networks.

The computational model predicts several pathways from MEK1 to c-Src based on protein-protein interaction data (see Fig. 3). Some of the biologically-relevant characteristics of the predictions include the prediction that all paths include SEK1 and p38 which have been shown to be downstream from MEK1 [17,10]. The fact that our method identified this biologically correct connectivity increases the confidence in the predicted pathways. Downstream from p38 there is a predicted bifurcation of signal with seven possible paths. However, these seven paths converge onto three molecules c-CBL, Caveolin1, and FADK1 which are directly upstream from c-Src.

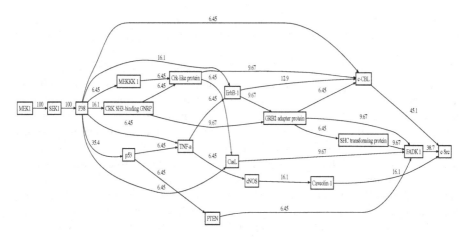

Fig. 3. A graphical representation of the paths predicted leading from MEK1 to c-Src. Each interaction (edges) is labeled by the % of paths that it appeared in. Since this is the percent of predicted
paths that pass through a given interaction, this number can be taken as an estimate of the importance of the interaction among all the interactions in the prediction. Note that this number should
not be confused with the confidence that the interaction exists—all interactions depicted in this
graph had support values greater than 99.9% as reported by the STRING database.

This modeling result is very interesting because it offers testable hypotheses to direct
the experimental validation of the predictions. The first experiment is to knock out
SEK1 or p38, anticipating that this would completely knock out connectivity between
MEK1 and c-Src. Experiments to inhibit the connectivity in this pathway would include
using siRNA to knock out expression of SEK1 and p38, and chemical intervention
experiment by using a pharmacological inhibitor of p38. If we experimentally observe
that, when p38 is inhibited, there is no change in connectivity between MEK1 and c-Src
this would direct us back to make changes in the model. If we observe only partial loss
of connectivity when p38 is blocked, this would imply additional pathways not utilizing
p38, and this again would direct us back to refine our model. Additionally, knocking out
or pharmacologically inhibiting c-CBL, Caveolin1, or FADK1 should give one of three
results complete, partial, or no loss of connectivity between MEK and c-Src. Based on
the results from these experiments we would be able to determine novel connectivity
between MEK1 and c-Src in a quick and directed manner. Therefore, by this modeling-
based hypothesis-driven method, coupled with targeted experimental manipulations, we
can rapidly identify novel connectivity between signaling molecules and pathways.

3   Materials and Methods

3.1   Knockout Experiment Design

In order to quantify changes in several nodes of the signaling network in parallel we
used the reverse phase protein micro-array technology. Using this proteomic tool we
were able to measure changes in the activity state as well as total levels of expressed
proteins. The method is described below.

Protein Lysate Micro Array. Arrays were prepared using cells lysed on ice with microarray lysis buffer (50 mM Hepes, 150 mM NaCl, 1mM EGTA, 10 mM Sodium Pyrophosphate, pH 7.4, 100 nM NaF, 1.5 mM MgCl2, 10% glycerol, 1% Triton X-100 plus protease inhibitors; aprotinin, bestatin, leupeptin, E-64, and pepstatin A). Cell lysates were centrifuged at 15,000 g for 10 minutes at 4C. Supernatant was collected and quantified using using a protein-assay system (Bio-Rad, Hercules, CA), with BSA as a standard. Using a GeneTac G3 DNA arrayer (Genomic Solutions, Ann Arbor, MI), six two-fold serial dilutions of cell lysates are arrayed on multiple nitrocellulose-coated glass slides (FAST Slides, Whatman Schleicher & Schuell, Keene, N.H). Arrays were produced in batches of 10. Printed slides were stored in dessicant at -20C. Antibodies were screened for specificity by Western blotting. An antibody was accepted only if it produced a single predominant band at the expected molecular weight. Each array was incubated with specific primary antibody, which was detected by using the catalyzed signal amplification (CSA) system (DAKO). Briefly, each slide was washed in a mild stripping solution of Re- Blot Plus (Chemicon International, Temecula, CA) then blocked with I- block (Tropix, Bedford, MA) for at least 30 minutes. Following the DAKO universal staining system, slides were then incubated with hydrogen peroxide, followed by Avidin for 5 minutes, and Biotin for 5 minutes. Slides were incubated with primary and secondary antibodies then incubated with streptavidin-peroxidase for 15 minutes, biotinyl tyramide (for amplification) for 15 minutes, and 3,3-diaminobenzidine tetrahydrochloride chromogen for 5 minutes. Between steps, the slide was washed with TBS-T buffer. Each slide was probed with validated antibodies under optimal blocking and binding conditions. Loading is determined by comparing phosphorylated and non-phosphorylated antibodies as well as by assessing control antibodies to prevalent and stable proteins. Six serial dilutions of each sample facilitate quantification and ensure that any slide can be assessed with different antibodies. Multiple controls are placed on each slide to facilitate quantification and robustness of the assay. Data are collected and analyzed by background correction and spot intensity using Image J. Protein phosphorylation levels are expressed as a ratio to equivalent total proteins. Fold increases in spot intensities were calculated against non-stimulated control samples. The following antibodies were used: EGFR, c-Src, Stat3, MAPK1,2, AKT, S6K, MEK1, NFkB, BAD, p38 MAPK, phosho c-Src, phospho Stat3, phospho AKT, phospho S6K, phospho MEK1, phospho NFkB, phospho BAD, phospho p38 MAPK.

3.2 Predicting Novel Pathways Based on Knockout Results

After completing the set of knockout experiments, we conducted a novel computational analysis to predict new pathways needed to explain the experimental results. This analysis consisted of two main stages:

1. *Identifying inconsistent results:* in this step we identified any individual knockout experiments that could not be explained by the model network. We call these results *inconsistent.*
2. *Constructing candidate pathways:* for each inconsistent result, we performed an exhaustive search of protein interaction data for hypothetical pathways that could explain the result and augment the existing incomplete model.

It is important to recall from Fig. 1 that there are multiple explanations for inconsistent results—only one of which is the incompleteness of the model. To be concrete, the experimental results presented in this paper can also be explained by undesired drug interactions with proteins other than MEK1. Our analysis finds several very viable pathways that may be missing from this network, making our approach valuable to the experimental biologist. However, in a complete analysis other sources of error must be taken into account. We identify these other sources of inconsistency as directions for future work, focusing in this paper only on the prediction of new pathways to handle the case of an incomplete model.

In the following sections we provide a detailed description of the steps itemized above.

Identifying Inconsistent Results. In order to determine which experimental results were unexpected, it was necessary to select a model signaling network that contained the complete set of known and relevant interactions. Since all of our experiments involved proteins embedded in the EGFR network, we used a model based on an extensive literature review of interactions in this network [11]. We stored the model signaling network as a pathway graph model [14]. In this representation, each protein/protein-state pair (e.g. AKT-inactive, AKT-active, and EGFR-phosphorylated) and each interaction is represented by a node. Directed edges connect protein/state pairs to interactions (reactions) they participate in and connect reactions to protein/state pairs that are produced as a result of the interaction. This representation explicitly depicts all experimentally derived and published paths through the signaling network—allowing extensive analysis of the connectivity within the network.

Recall that a knockout or inhibition result can indicate that a signaling pathway exists between two proteins (as was the case with *MEK1* and *c-Src* in the experiments described above). When a knockout or inhibition experiment yields such a result for proteins X and Y, but no chain of directed interactions exists in the model network between X and Y, we call this result *inconsistent*—implying that the model is not capable of explaining the result and requires the addition of a new pathway.

In order to identify inconsistent results, we first selected only those results which indicated the presence of a signaling pathway between two proteins. For each of these results, we used the constrained downstream algorithm [14] to enumerate all paths between the two proteins in the model. This algorithm performs an exhaustive search of a pathway graph model for all paths connecting one set of proteins to another. In this algorithm, the first protein is considered the source, the second protein is considered the sink, and all paths found are directed from the sources to sinks, as they would occur in the signaling network.

For the experiments we considered for this paper, the downstream algorithm reported paths for all results except *MEK1* to *c-Src*. The absence of any path from *MEK1* to *c-Src* indicates that the model cannot explain the inhibitory result observed between these two proteins. As a result, this result was identified as an inconsistent result.

Constructing Candidate Pathways. In this step, given an inconsistent result, we seek a set of candidate pathways, any of which can explain the result observed. For the inconsistent result supporting a pathway between proteins X and Y, we know that the model has insufficient interactions to connect them. Therefore, we must look elsewhere in order to find biologically-relevant interactions to connect these two proteins.

Protein interaction databases are, effectively, massive repositories of putative protein interactions. Despite the fact that many of the interactions may not, in reality, occur, these databases provide a good source of interactions to use when assembling hypothetical pathways.

One issue that must be addressed is the fact that many studies have shown the interactions in these databases to be of varying quality [4,2]. Since we seek biologically-likely pathways which are, by definition, composed of biologically likely interactions, we must have some way of evaluating the *confidence* of any given interaction in the database. Significant work has been done into the problem of assigning confidence to interactions [9,4,2,18,16]. In this study, we made use of the STRING database [19] which provides interactions with confidence scores. However, using other interaction databases and other confidence scoring schemes are equally valid approaches and, depending on the interactions in the database and how confidence is estimated, may produce somewhat different results from ours.

Once a protein interaction database and confidence scoring scheme have been selected, a protein interaction network can be constructed. This is a data structure that combines the interactions in the database with the scoring scheme. In this network, a node is a protein, an edge $e = (u, v)$ is an undirected interaction between proteins u and v. Each edge, $e = (u, v)$ is assigned a weight equal to its log-likelihood score: $weight(e) = -log(c(e))$, where $c(e)$ is the confidence assigned to interaction e by the scoring scheme.

When constructed as described, this network has the special property that the weight of path $\langle u_1, u_2, ..., u_n \rangle$ within this network has the following correspondence to its total support:

$$\sum_{i=1}^{n-1} w((u_i, u_{i+1})) = -log(\prod_{i=1}^{n-1} c((u_i, u_{i+1}))).$$

Since the function $-log(x)$ approaches 0 as $x \to 1$, the sum on the left will be smallest when the individual path edges have confidence scores closest to 1. Therefore, the shortest (lightest) path in the network between nodes X and Y corresponds to the most biologically-likely pathway connecting the two proteins represented by nodes X and Y.

Since all paths within some confidence threshold probably correspond to some biologically-likely pathway, we choose to search for the set of k-shortest paths—where k is a parameter indicating how many paths we want to retrieve. Paths should be reported in order of increasing weight so that the kth path is the longest (least biologically-likely) of the paths returned by the search.

Significant work has been done on the problem of enumerating the k-shortest paths and efficient algorithms exist for solving it [5,1]. For our purposes in this project, we use a variant of the k-shortest path problem, called the k-shortest *simple* path problem [22,13]. A simple path is one that contains no loops. The reason for this restriction is that, while feedback loops are quite common in signaling pathways, we are only interested in the simplest pathways that can explain the inconsistent results. Under the log-likelihood transformation, edges with 100% support will have zero weight, creating the possibility of cycles in the graph. As a result, we choose to discard any short paths that contain loops from the set of candidate pathways.

In our analysis, we used an implementation of Eppstein's k-shortest paths algorithm [5]. Non-simple paths were detected and removed from the output in order to give a k-shortest simple paths algorithm. We ran the algorithm and found the 100 shortest simple paths. A detailed analysis of these paths is given in Section 2.2.

As a final step in identifying the candidate pathways, direction must be imposed on the paths extracted. The paths extracted from the protein interaction network are bi-directional since the edges are undirected. For a result in which a knocking out protein X caused a change in protein Y, the pathway direction is towards protein Y. In order to model this in the interaction network, we always search for paths from X to Y and report the the nodes of each path in the order in which they appear—from first to last.

3.3   The PathwayOracle Tool

In the past ten to fifteen years biologists have uncovered hundreds of interactions within signaling pathways in biological systems. A challenge given this large amount of data is to develop novel methods to probe the data and ask questions that cannot be answered by experimental biology alone. On the other hand it is also vital to integrate the experimental biology with the computational models and methods.

In order to address these issues, we have created the PathwayOracle software package which contains various tools enabling the computational analysis and extension of experimental results and techniques [14]. The novel approach to pathway prediction described in this paper is the most recent addition to the PathwayOracle package. Included with the implementation is the human subset of the interactions in the STRING database, though other interaction datasets can be specified.

The entire toolkit is open-source, implemented in Java, and available upon request from the authors. Additional information about other features and tools included in the package is available on the website:
`http://bioinfo.cs.rice.edu/pathwayoracle`.

Acknowledgments

We gratefully acknowledge the help that Melissa Muller provided in assembling the figures for the experimental results. We appreciate Erion Plaku's time and effort spent debugging an implementation of the shortest path algorithm.

This work was supported in part by DOD grant BC044268 to PTR.

References

1. Ahuja, R., Mehlhorn, K., Tarjan, B.: Faster Algorithms for the Shortest Path Problem. Journal of Association of Computing Machinery 37, 213–223 (1990)
2. Bader, J., Chaudhuri, A., Rothberg, J., Chant, J.: Gaining confidence in high-throughput protein interaction networks. Nature Biotechnology 22(1), 78–85 (2004)
3. Belloni, E., Muenke, M., Roessier, E., Traverse, G., Siegel-Bartelt, J., Frumkin, A., Mitchell, H.F., Donis-Keller, H., Helms, C., Hing, A.V., Heng, H.H.Q., Koop, B., Martindale, D., Rommens, J.M., Tsui, L.-C., Scherer, S.W.: Identification of Sonic hedgehog as a candidate gene responsible for holopro-sencephaly. Nature Genetics 14, 353–356 (1996)

4. Deng, M., Sun, F., Chen, T.: Assessment of the reliability of protein-protein interactions and protein function prediction. In: Proceedings of the Eight Pacific Symposium on Biocomputing, pp. 140–151 (2003)

5. Eppstein, D.: Finding the k shortest paths. SIAM Journal of Computing 28(2), 652–673 (1998)

6. Feldman, D.S., Carnes, C.A., Abraham, W.T., Bristow, M.R.: Mechanisms of Disease: β-adrenergic receptors alterations in signal transduction and pharmacogenomics in heart failure. Nature Clinical Practice Cardiovascular Medicine 2, 475–483 (2005)

7. Hennessy, B.T., Smith, D.L., Ram, P.T., Lu, Y., Mills, G.B.: Exploiting the PI3K-AKT pathway for cancer drug discovery. Nature Review Drug Discovery 4, 988–1004 (2005)

8. Hunter, T.: Signaling – 2000 and beyond. Cell 100(1), 113–127 (2000)

9. Hwang, D., Rust, A.G., Ramsey, S., Smith, J.J., Leslie, D.M., Weston, A.D., Atauri, P., Aitchison, J.D., Hood, L., Siegel, A.F., Bolouri, H.: A data integration methodology for systems biology. PNAS 102(48), 17296–17301 (2005)

10. Johnson, G.L., Lapadat, R.: Mitogen-activated protein kinase pathways mediated by ERK, JNK, and protein kinases. Science 1912, 38 (2002)

11. Oda, K., Matsuoka, Y., Funahashi, A., Kitano, H.: A comprehensive pathway map of epidermal growth factor signaling. Molecular Systems Biology, msb41000014–E1–E17 (2005)

12. Pagel, P., Kovac, S., Oesterheld, M., Brauner, B., Dunger-Kalthenbach, I., Frishman, G., Montrone, C., Mark, P., Stumpflen, V., Mewes, H., Reupp, A., Frishman, D.: The MIPS mammalian protein-protein interaction database. Bioinformatics 21(6), 832–834 (2005)

13. Pascoal, M., Martins, E.: A new implementation of Yen's ranking loopless paths algorithm. Quarterly Journal of the Belgian, French, and Italian Operations Research Societies 1(2), 121–134 (2003)

14. Ruths, D., Nakhleh, L., Iyengar, M.S., Reddy, S.A.G., Ram, P.T.: Graph-theoretic Hypothesis Generation in Biological Signaling Networks. Journal of Computational Biology 13(9), 1546–1557 (2006)

15. Scott, J., Ideker, T., Karp, R.M., Sharan, R.: Efficient Algorithms for Detecting Signaling Pathways in Protein Interaction Networks. In: McLysaght, A., Huson, D.H. (eds.) RECOMB 2005. LNCS (LNBI), vol. 3678, pp. 1–13. Springer, Heidelberg (2005)

16. Sharan, R., Suthram, S., Kelly, R.M., Kuhn, T., McCuine, S., Uetz, P., Sittler, T., Karp, R.M., Ideker, T.: Conserved patterns of protein interaction in multiple species. PNAS 102(6), 1974–1979 (2005)

17. Uhlik, M.T., Abell, A.N., Cuevas, B.D., Nakamura, K., Johnson, G.L.: Wiring diagrams of MAPK regulation by MEKK1, 2, and 3. Biochem. Cell Biol. 82(6), 658–663 (2004)

18. von Mering, C., Krause, R., Snel, B., Cornell, M., Oliver, S.G., Fields, S., Bork, P.: Comparative assessment of large-scale data sets of protein-protein interactions. Nature 417, 399–403 (2002)

19. von Mering, C., Heynen, M., Jaeggi, D., Schmidt, S., Bork, P., Snel, B.: STRING: a database of predicted functional associations between proteins. Nucleic Acids Research 31(1), 258–261 (2003)

20. Yeang, C.H., Ideker, T., Jaakkola, T.: Physical network models. Journal of Computational Biology 11, 243–262 (2004)

21. Yeang, C.H., Mak, H.C., McCuine, S., Workman, C., Jaakkola, T., Ideker, T.: Validation and refinement of gene regulatory pathways on a network of physical interactions. Genome Biology 6(7), R62 (2005)

22. Yen, J.Y.: Finding the K Shortest Loopless Paths in a Network. Management Science 17(11), 712–716 (1971)

Alignment of Mass Spectrometry Data by Clique Finding and Optimization

Daniel Fasulo[1], Anne-Katrin Emde[1], Lu-Yong Wang[1], Karin Noy[1],
and Nathan Edwards[2]

[1] Integrated Data Systems Department
Siemens Corporate Research,
755 College Road East, Princeton, NJ, USA
[2] Center for Bioinformatics and Computational Biology
3119 Biomolecular Sciences Bldg. #296
University of Maryland
College Park, MD 20742

Abstract. Mass spectrometry (MS) is becoming a popular approach for quantifying the protein composition of complex samples. A great challenge for comparative proteomic profiling is to match corresponding peptide features from different experiments to ensure that the same protein intensities are correctly identified. Multi-dimensional data acquisition from liquid-chromatography mass spectrometry (LC-MS) makes the alignment problem harder. We propose a general paradigm for aligning peptide features using a bounded error model. Our method is tolerant of imperfect measurements, missing peaks, and extraneous peaks. It can handle an arbitrary number of dimensions of separation, and is very fast in practice even for large data sets. Finally, its parameters are intuitive and we describe a heuristic for estimating them automatically. We demonstrate results on single- and multi-dimensional data.

Keywords: mass spectrometry, alignment, bounded error model, clique finding.

1   Introduction

Mass spectrometry (MS) and liquid-chromatography coupled with mass spectrometry (LC/MS) have increasingly become the methods of choice for analysis of complex protein mixtures as advances in technology have enabled the routine study of large biomolecules. These techniques have demonstrated the capability to discover potential biomarkers and therapeutic targets, and form the basis of new diagnostics molecular diagnosis[1] [2]. The analysis and interpretation of the enormous volumes of proteomic data remains a demanding challenge [3].

In this paper, we focused on the spectral alignment problem, in which the task is to match corresponding peptide features from different experiments to ensure that the same protein intensities are correctly identified. Unfortunately, few investigators recognize the importance of this problem in obtaining high-quality results from proteomics investigations. Recently Semmes, *et al.*, utilized data collected from different spectrometers in different physical locations. They conclude that solving the alignment problem

T. Ideker and V. Bafna (Eds.): Syst. Biol. and Comput. Proteomics Ws, LNBI 4532, pp. 119–129, 2007.

is critical to ensuring data compatiblilty and reproducibility [4]. Mass spectrometry proteomics is becoming more popular and more common [5], and the alignment problem becomes more demanding when different machines are used to generate spectra or spectra are generated over a long time period. The higher dimensional protocols, such as variations on LC/MS, makes the problem more imperative[2].

Here, we propose a general peak alignment method called BAG, and demonstrate its utility on single- (SELDI-TOF) and multi-dimensional data. We present our method as a general alignment framework that may be specialized for many experimental protocols employing spectrometry and chromatography, including those employing multi-dimensional separations.

2   Methods

We propose a general framework for aligning peaks based on a bounded error model. The peaks may be single- or multi-dimensional. Our paradigm is similar in spirit to certain clustering methods, but is specialized to a data model in which there are hard constraints on the way in which objects can be grouped. The constraints are designed so that an algorithm can efficiently describe all possible solutions that obey the constraints. An optimization method is then used to select the best solution. The number of possible solutions under the constraints is typically much smaller than all possible partitions of the data, making this method efficient.

Our method is quite general, so we present it in terms of experiments which measure properties of anonymous objects. Our goal is to determine, based solely on the measurements, which objects (*e.g.* "peaks") are identical across the set of experiments.

2.1   The Model

In our model, an *object* refers to an object (*e.g.* a peptide) to be observed in some experiments through a set of d sensors. Let $\mathfrak{U} = \{\mathcal{O}_1, \mathcal{O}_2, \ldots, \mathcal{O}_M\}$ denote the universe of possible objects of interest in the experiments.

Our experiments are imperfect in two senses. First, we assume that the experiments are processed independently through an imperfect method to detect the objects. Let $\mathcal{E} = \{E_1, E_2, \ldots, E_K\}$ be the set of K experiments. Each experiment E_k is itself a set of *features*,i.e., observed peaks, denoted $f_k^{(i)}$ where $1 \leq i \leq |E_k|$, that have been detected and are believed to correspond to objects. However, some features may be spurious (false positives); in other cases no feature is detected (false negatives, or absence from particular experiments). In the case where the feature $f_k^{(i)}$ does correspond to an object $\mathcal{O}_m$, we define $\pi(f_k^{(i)}) = \{\mathcal{O}_m\}$; otherwise, $\pi(f_k^{(i)}) = \emptyset$. Second, we model the features as being subject to measurement error. Each feature $f_k^{(i)}$ is represented as a vector in $\mathbb{R}^d$, whose entries correspond to readings from the d sensors. The reading from sensor j is denoted $f_k^{(i)}[j]$.

The following key constraint is imposed. Let $\epsilon_j : \mathbb{R} \rightarrow \{\mathbb{R}\}$ be an error bound on the features in dimension j, so ϵ_j maps a feature f to an interval $[\epsilon_j^\ell(f[j]), \epsilon_j^r(f[j])]$.

Definition 1 (Bounded Measurement Error). *A data set has bounded measurement error if we can select ϵ_j for all $1 \leq j \leq d$ such that for all features $f_1, f_2 \in \bigcup_{k=1}^{K} E_k$ satisfying $\pi(f_1) \cap \pi(f_2) \neq \emptyset$, the following holds: $\epsilon_j(f_1[j]) \cap \epsilon_j(f_2[j]) \neq \emptyset$.*

We select the bounded measurement error model to correspond to the standard wet lab convention that, for example, mass spectrometers are accurate to plus or minus some percentage, and that chromatography peaks are reproduced within an interval of plus or minus some number of seconds. We also note that mathematically, the bounded measurement error model must apply to any retrospective analysis of a finite data set, although the bounds may not be known *a priori*.

Under a parsimony assumption, two features correspond observations of the same object if the intervals formed by error bounds associated with each corresponding sensor measurement intersect. Two features which satisfy these constraints are defined as *compatible*. Given a set of K experiments, our goal is to partition the set of all features into subsets which correspond to the same underlying object. A three-step procedures will be described.

2.2 Box Creation and Parameter Estimation

We create a box for each feature by constructing the constraint intervals in each of the d dimensions. In the case where there is little or no prior information on the appropriate interval widths, a heuristic approach can be utilized.

Our heuristic is based on some assumptions about the error bounds and the box overlap graph: First, each ϵ_j is parameterized by a single numerical parameter θ_j, $\theta_j \geq 0$. The application of the function under this parameter on feature f is denoted $\epsilon_j(f[j], \theta_j)$, and we let θ be the vector of θ_j, $1 \leq j \leq d$. Second, the number of overlaps in $\mathcal{F}$ in dimension j induced by ϵ_j increases monotonically with θ_j. Furthermore, many objects are assumed to be observed in all experiments, as one would expect in most analysis of complex biological mixtures. If these assumptions are valid, then we can choose the vector θ that induces a set of boxes $\mathcal{B}$ and overlap graph $G(\mathcal{B})$ where the number of connected components in $G(\mathcal{B})$ that are complete subgraphs of size K is maximized.

Let Ω represent the universe of all possible choices of θ. Since $\theta \in \mathbb{R}^d$, Ω may initially appear to be infinite. However, since our metric depends on the finite number of possible configurations of $G(\mathcal{B})$, we will show that Ω can effectively be represented by a finite set of vectors whose values induce the different configurations.

Let f and f' be features, and let $\hat{\theta}(f, f')$ be such that

$$\hat{\theta}_j(f, f') = \min_{\theta_j \geq 0} \left(\epsilon_j(f, \theta_j) \cap \epsilon_j(f', \theta_j) \neq \emptyset \right),$$

i.e., $\hat{\theta}_j(f, f')$ represents the smallest value of θ_j such that f and f' are compatible. Now consider features f'' and f'''. We say that $\hat{\theta}(f, f') \preceq \hat{\theta}(f'', f''')$ if and only if $\hat{\theta}_j(f, f') \leq \hat{\theta}_j(f'', f''')$ for all j, $1 \leq j \leq d$. Note that this relationship implies that under parameters $\hat{\theta}(f'', f''')$, f and f' are also compatible.

Let $\hat{\Omega}$ be the set $\left\{ \hat{\theta}(f, f') : f, f' \in \mathcal{F} \right\}$. Now define define the $\oplus$ operator such that $\theta \oplus \theta'$ is a d-dimensional vector where element j, $1 \leq j \leq d$, is defined as $\max\{\theta_j, \theta'_j\}$.

Let Ω_C denote the closure of $\hat{\Omega}$ under $\oplus$. The theorem below (proof omitted) shows that Ω can effectively be represented by Ω_C.

Theorem 1. *Let $\mathcal{B}'$ be the set of boxes derived from $\mathcal{F}$ by $\theta' \in \Omega$ and $G(\mathcal{B}')$ be the overlap graph derived from $\mathcal{B}'$. There there exists a $\theta \in \Omega_C$ which derives a set of boxes $\mathcal{B}$ from $\mathcal{F}$ such that $G(\mathcal{B})$ is identical to $G(\mathcal{B}')$.*

Let θ^* be a vector such that for all j, $1 \leq j \leq d$, $\theta_j^* = \theta_j$ for some $\theta \in \hat{\Omega}$. Let $\Omega^* = \{\theta^*\}$ given $\hat{\Omega}$. It can be shown that $\Omega_C \subseteq \Omega^*$ (proof omitted).

It is simple to enumerate the elements of Ω^*. The enumeration process can be accelerated by noting that if $\theta \preceq \theta'$ and $G(\mathcal{B})$ and $G(\mathcal{B}')$ are the respective graphs induced by θ and θ', then $G_E(\mathcal{B}) \subseteq G_E(\mathcal{B}')$. The standard UNION-FIND data structure can thus be used to identify the connected components of $G(\mathcal{B})$ as edges are added. Since the size of Ω^* is $O(n^{2d})$, a faster heuristic method is needed for most applications. We recommend a steepest descent method in which the state θ has successors $\{\theta' | \theta \preceq \theta'\}$. Various heuristics can be used to choose the initial state and to skip evaluation of neighboring states that are derived from negligible changes in parameter values.

2.3   Maximal Clique Finding

This section describes a method to find all maximal sets of mutually compatible features. Given the error function $\epsilon_j (1 \leq j \leq d)$, we can convert our features into boxes in d-dimensional space. Let $\mathcal{B}$ be a set of n iso-oriented boxes in $\mathbb{R}^d$. Each box $B_i \in \mathcal{B}$ can be represented as a set of d non-empty intervals, each denoted $X_j(B_i)$, $1 \leq j \leq d$, and called the *extent* of B_i in dimension j.

We use the term *clique* to refer to a set of mutually intersecting boxes. It is easy to show that such sets have the Helly property; that is, if $C = \{B_1, B_2, \ldots, B_m\}$ is a clique, then $\forall j$, $1 \leq j \leq d$, $\bigcap_{i=1}^m X_j(B_i) \neq \emptyset$. We denote the area of intersection for clique C as box A_C and borrow corresponding notation to say that the extent of A_C in dimension j is $X_j(A_C)$, where

$$X_j(A_C) = [x_j^\ell(A_C), x_j^r(A_C)] = \bigcap_{B \in C} X_j(B)$$

Our goal, given the set $\mathcal{B}$, is to explicitly find all maximal cliques occurring in $\mathcal{B}$.

To describe our solution, let $G(\mathcal{B})$ be an undirected graph such that there is a vertex corresponding to each box in $\mathcal{B}$ and an edge between every pair of intersecting boxes. Such a graph is called the *box intersection graph*, and there is an obvious correspondence between the maximal cliques in this graph and the maximal cliques defined.

For $d > 1$, we define the *slice* operator on box B at x, $S^d(B, x)$, as the projection of B into $\mathbb{R}^{d-1}$ obtained by dropping X_d if $x \in X_d$, or $\emptyset$ otherwise. We define the *slice set* of box B_i, $\mathcal{S}_i^d$, as follows:

$$\mathcal{S}_i^d = \left\{ S^d(B_j, x_d^r(B_i)) : S^d(B_j, x_d^r(B_i)) \neq \emptyset, 1 \leq j \leq n \right\}.$$

We now use the slice set concept and a small set of theorems to propose a recursive method for finding the maximal cliques of $\mathcal{B}$. The recursion proceeds on the number of dimensions, and the base case is reached when $d = 1$ (or, optionally, when $d = 2$).

Theorem 2. *Let C be a maximal clique of $G(\mathcal{B})$. Then C is a maximal clique of $G(\mathcal{S}_i^d)$ for some $B_i \in C$.*

Proof. Let $B_i \in C$ be the box with minimum $x_d^r(B_i)$. Since C is a clique, it must be the case that for all $B \in C$, $x_d^\ell(B) < x_d^r(B_i)$. Furthermore, by definition, $x_d^r(B) \geq x_d^r(B_i)$. Therefore, all elements of C occur in $\mathcal{S}_i^d$. It is easy to see that by the definition of a clique, all elements of a clique in $\mathbb{R}^d$ must form a clique in their first $d - 1$ dimensions; hence, the elements of C form a clique in $\mathcal{S}_i^d$. Finally, C must be maximal with respect to $\mathcal{S}_i^d$. If there were some other box B' that were in $\mathcal{S}_i^d$ and could be added to C, then this rectangle would also intersect all rectangles in dimension d at $x_d^r(B_i)$ and hence C would not be maximal in $G(\mathcal{B})$.

We denote the set of maximal cliques of $G(\mathcal{B})$ as $\mathcal{C}$, and the set of maximal cliques in $G(\mathcal{S}_i^d)$ that contain B_i as $\mathcal{C}_i^d$. The consequence of Theorem 2, stated succinctly as $\mathcal{C} \subseteq \bigcup_{i=1}^n \mathcal{C}_i^d$, shows how we might proceed toward finding the maximal cliques of $\mathcal{B}$:

Step 1. If $d = 1$, calculate the maximal cliques of $\mathcal{B}$ directly.
Step 2. Otherwise, calculate each $\mathcal{S}_i^d$ and recursively find the corresponding $\mathcal{C}_i^d$.
Step 3. Filter out those elements of $\mathcal{C}_i^d$ which are not maximal with respect to $G(\mathcal{B})$.

A simple sweepline procedure for Step 1 was first described by Turnbull [6]. Let $\mathcal{I}$ be a set of n intervals in $\mathbb{R}$. For each $I_i \in \mathcal{I}$, let $I_i = [x^\ell(I_i), x^r(I_i)]$. As before, for simplicity of presentation we assume all of the interval end points are unique. Let P be the set of all interval end points; that is, $P = \bigcup_{i=1}^n \{x^\ell(I_i), x^r(I_i)\}$. Let $\boldsymbol{P}$ be a vector of length $2n$ containing the elements of P sorted in increasing order. Finally, let $p(i) = \ell$ if $\boldsymbol{P}[i] = x^\ell(I_j)$ for some j; otherwise $p(i) = r$ because it must be the case that $\boldsymbol{P}[i] = x^r(I_j)$ for some j.

Let $\mathcal{S}_i$ denote the set of intervals containing the point $\boldsymbol{P}[i]$. For completeness, define $p(0) = r$. Turnbull proved the following theorem (albeit with different notation):

Theorem 3 (Turnbull 1976). *$\mathcal{S}_i$ is a maximal clique of intervals if and only if $p(i) = r$ and $p(i - 1) = \ell$.*

If $d = 2$, we can also adopt a direct algorithm by Maathuis [7] ($O(n^2)$ time) instead of the recursive step. Since each $\mathcal{S}_i^d$ is simply a set of boxes in $\mathbb{R}^{d-1}$, Step 2 is a straightforward recursive usage of the algorithm. The subtle catch is that only those cliques containing B_i are retained for Step 3. Step 3 depends on the construction and processing of the slice sets in a particular order. Let P_d be the set of all interval end points in dimension d; that is, $P_d = \bigcup_{i=1}^n \{x_d^\ell(B_i), x_d^r(B_i)\}$. Let $\boldsymbol{P}_d$ be a vector of length $2n$ containing the elements of P_d sorted in increasing order.

Let L be a data structure representing a set of boxes, *e.g.*, a hash table that uses the index i of each B_i as a key. We enumerate the slice sets by considering each member x of $\boldsymbol{P}_d$ in increasing order. There are two cases for each x: either $x = x_d^\ell(B_i)$ or $x = x_d^r(B_i)$ for some $B_i \in \mathcal{B}$, meaning that either x is the start of B_i in a left-to-right sweep of dimension d, or the end. First, suppose x is the start of B_i. In this case, we insert B_i into L. If x is the end B_i, then L contains exactly those intervals in $\mathcal{S}_i^d$. We extract $\mathcal{S}_i^d$, remove B_i from L, and recursively process $\mathcal{S}_i^d$ to generate $\mathcal{C}_i^d$. The following theorem demonstrates why it is useful to generate and process the slice sets in this order.

Theorem 4. *Let $C' \in \mathcal{C}_j^d$ be a maximal clique of $G(\mathcal{S}_j^d)$ that is not maximal with respect to $G(\mathcal{B})$. Then there exists a clique $C \in \mathcal{C}_i^d$ with $x_d^r(B_i) < x_d^r(B_j)$ such that $C' \subseteq C$.*

Proof. By Theorem 2, a maximal clique C of $G(\mathcal{B})$ that contains C' *must* be contained in some $\mathcal{C}_i^d$. Suppose that $x_d^r(B_i) > x_d^r(B_j)$. By definition $B_j \in C'$, but $B_j \notin C$ because $S^d(B_j, x_d^r(B_i))$ must be $\emptyset$. The implication is that $C' \nsubseteq C$. Hence, it must be the case that $x_d^r(B_i) < x_d^r(B_j)$.

Thus, if we consider the cliques in increasing order of x_d^r, we can guarantee that all cliques found in the slice sets that are not maximal with respect to $G(\mathcal{B})$ will be observed after their containing maximal clique. A sufficiently fast way to check whether a clique is a sub-clique of the previously-observed maximal cliques is critical. Testing for clique containment is accomplished via the next theorem.

Theorem 5. *Let $C \in \mathcal{C}$ be a maximal clique of $G(\mathcal{B})$. The clique $C' \subseteq C$ if and only if $A_C \subseteq A_{C'}$.*

Proof. Suppose first that $C' \subseteq C$. It follows immediately from the comments in Section 2.3 on areas of intersection that $A_C \subseteq A_{C'}$. Therefore, the centroid of A_C is contained in $A_{C'}$. Conversely, if $A_C \subseteq A_{C'}$ and let x be an arbitrary point such that $x \in A_C$. This implies $x \in A_{C'}$, so all of the rectangles of C' must also contain x. Hence all rectangles in C and C' share a common point of intersection, so the set $C'' = C \cup C'$ is a clique. Since C is maximal, this means that $C'' \subseteq C$, and hence $C' \subseteq C$.

Thus, to test if C' is a sub-clique of a previously-observed clique C, we can select an arbitrary point x (*e.g.* centroid of A_C) from each clique C. This creates a set of points $\bar{X}$. When then considering a subsequently-detected clique C', we observe that C' is maximal if and only if $\bar{X} \cap A_{C'} = \emptyset$. Hence, we only need to test if some point in $\bar{X}$ is contained in a box $A_{C'}$, a problem for which efficient solutions exist.

2.4   Resolving Ambiguities by Constrained Optimization

These sets satisfy the constraint imposed by the bounded measurement error model, but they do not completely solve the problem of partitioning the input into sets of identical objects. Violations of transitivity cause features to appear in multiple sets; additionally, features from the same set may not be identical when other evidence is examined. Here, we place the formation of the final partition into the context of an optimization problem to solve possible ambiguities.

Let $\mathcal{F}$ denote the set of features across all experiments. We are only concerned with "true" feature set $\mathcal{F}_\pi$ such that $\pi(f) \neq \emptyset$. An identity relationship partitions $\mathcal{F}_\pi$ into equivalence classes $\Pi_1, \Pi_2, \ldots, \Pi_R$. We are interested in finding the partition which satisfied the relationship $f_1, f_2 \in \Pi_r$ if and only if $\pi(f_1) = \pi(f_2)$. We assume that we have access to properties of the feature $\pi'(f)$ and we construct a function $\varphi(f, f')$ that approximates $\Pr[\pi(f) = \pi(f')|\pi'(f), \pi'(f')]$. Given such a function, we can attempt to find a partition of π that maximizes

$$\left(\prod_{r=1}^{R} \prod_{f,f' \in \Pi_r} \varphi(f, f') \right) \cdot \left(\prod_{r=1}^{R} \prod_{f \in \Pi_r, f' \notin \Pi_r} (1 - \varphi(f, f')) \right) \qquad (1)$$

In general φ may be expensive to compute, and the search space of all possible partitions is large. We can restrict both the search space and the number of times φ is evaluated by using constraints imposed by the set of maximal cliques C found by the method of the previous section as shown below (proof omitted):

Theorem 6. *The partition of $\mathcal{F}_\pi$ which maximizes (1) satisfies the property that for all Π_r, $\Pi_r \subseteq C$ for some maximal clique $C \in \mathcal{C}$.*

Note that C is not a partition of $\mathcal{F}$ only because some features appear in more than one member of C; each feature is guaranteed to participate in at least one clique. Hence, it is possible to transform C into the optimal partition by performing a series of 2 operations. The first is *Assignment*: any feature which appears in multiple cliques must be assigned to a single clique and removed from the others. The second is *Partitioning* of cliques. In our implementation, simulated annealing[8] is used to search for the optimal partition, beginning with C and using the operations above to generate potential solutions.

3 Results

We tested our Java implementation of the BAG method on two data sets. The first set is composed of simulated two, three, and four dimensional data representing multidimensional proteomics data. The number of experiments ranges from 3 to 20, the number of features per experiment from 500 to 5000. The second set are real mass spectra acquired from a Ciphergen PBS2 SELDI-TOF mass spectrometer.

As a first step, the optimal parameters for the error bounds were computed for each set. The steepest descent method we used was able to find the optimal solution for all of the simulated sets. As the real MS data set is only one dimensional, the steepest descent in this case is a search of the whole parameter space and therefore the optimal solution must be found.

3.1 Alignment Accuracy

Based on the reference maps from which the simulated data sets were derived, the results were evaluated in terms of completeness and perfectness of the predicted groupings. Results for all subsets containing 1000, 2000 and 3000 true features per experiment are shown in Figure 1.

A clique is considered perfect if all features in this clique are derived from the same reference feature, and if there is a feature from each experiment present. In none of the sets the fraction of perfect cliques fell below 90%. Even with a higher number of noise points, such as 8000 in 8 maps, about 95% of the cliques were found to be perfect. The number of perfect cliques decreases with an increasing number of noise points. Therefore we introduced a second measure, namely completeness. In order to be a complete clique, the clique must contain all of the features associated with a single reference feature, and no features derived from another reference feature. It may contain additional noise points. In all of the sets, between 98.2% and 100% of the possible cliques were complete. Note that in our simulated data, our definition of φ depended

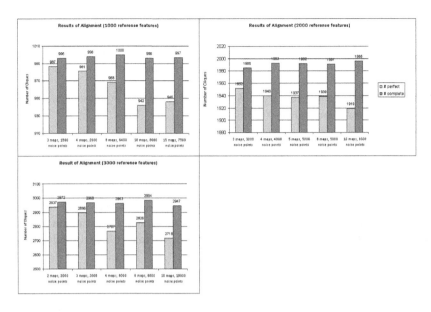

Fig. 1. Sets of 1000, 2000 and 3000 reference features per map were simulated, processed with the BAG program and then evaluated. The number of maps and noise points were varied.

only on the area of rectangle intersection. In real data, other criteria might be used to separate noise from the groups.

Despite the fact that our algorithm has a worst case running time that is exponential in the number of dimensions, we observe good performance on these data sets. Running times are less than one minute, and the rate of increase with the dimensionality appears linear on our simulated data (results not shown).

In a second study, we aligned a set of 50 SELDI-MS spectra containing total 7000 peaks. The data was derived from two subpopulations, which we denote A and B. For comparison, we also generated an alignment resulting from binning of the spectra (a commonly employed, simple heuristic); that is, partitioning the mass-charge ratio axis into fixed width bins, summing the peak intensities within each bin for each spectrum, and comparing bin values across spectra. We used equally spaced bins of 3000ppm width.

We computed Pearson correlation of the matched intensities between each pair of spectra and an associated p-value. For each spectrum we then calculated the mean of all correlation coefficients. As shown in Figure 2 the probability for the BAG data that there is no correlation is less than 0.05 for 48 of the 50 spectra. The binned results on the other hand show p values greater than 0.05 for almost half of the spectra. Using the BAG matrix the histogram of the mean Pearson correlation coefficients is clearly shifted to the right. The percentage of matched peaks/non-zero bins is only slightly higher for the BAG results. This indicates that the quality of the result is not only a matter of how many peaks/bins can be matched but primarily of how cleverly the matches are chosen.

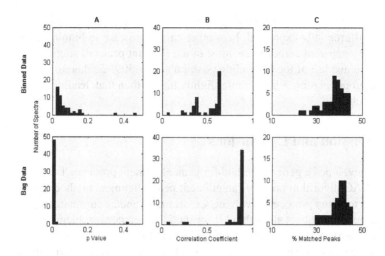

Fig. 2. Comparison of Binned matrix and BAG matrix. Values are means for each spectrum over all spectra. A. Test of the hypothesis of no correlation against the alternative that there is a correlation greater than zero. B. Pearson's Correlation Coefficient. C. Percentage of matched peaks/bins.

Using the Student's t-test for feature selection, the top discriminatory features between subpopulations A and B were extracted from each matrix. We found that selection of 10 features provided the best discriminatory power. We then assessed the effectiveness of k-nearest neighbor classification to distinguish between groups A and B in the resulting submatrices by leave-one-out cross-validation.

Table 1. The five features with the highest $-log$ p-values for each matrix

	Binned Data		Bag Data	
	m/z	$-log$ p-value	m/z	$-log$ p-value
1	5093.54	3.21283	4451.57	3.16409
2	6005.8	1.87027	6000.0	2.42041
3	3855.07	1.77192	1873.7	2.10496
4	1631.79	1.75043	1812.67	2.07766
5	1878.49	1.69186	1584.07	1.97849

A substantial improvement of classification accuracy is shown. The matrix formed with the 10 most discriminatory features from the BAG matrix yielded an overall 76% classification accuracy (80% sensitivity, 68% specificity). The matrix formed from the Binned matrix yielded a 66% classification accuracy (87% sensitivity, 32% specificity). We made the unsettling observation that the two methods identified very different features as being significantly different in the two subpopulations (see Table 1), so we investigated further. In one case where BAG detected a differential feature and binning did not, two peaks of subpopulation A and one peak of subpopulation B were assigned to the wrong bin. This relatively slight mistake prevented the feature from being identified as differentially expressed. In another case, a mistake in binning created a false positive. By apparent coincidence, most of the relevant peaks of subpopulation B happened to be just left of the bin ending position at m/z 5093.5 whereas almost all of the peaks of subpopulation A had an m/z slightly higher than that, leading to an artificially high p-value.

4   Discussion and Conclusions

We have developed a general method for peak alignment problem. The results indicate that our BAG algorithm provides an efficient peak alignment method that is generally applicable to many protocols involving spectrometry and/or chromatography.

We adopted Maathuis algorithm [7] for the 2D alignment problem, as an alternative to the general recursive method. This result is the most recent in among other papers for finding cliques in rectangle overlap graphs [9][10][11]. These algorithms derive from the problem of non-parametric maximum likelihood estimation using a box kernel, a statistical approach for estimating a phenonenon based on a number of observations. A related approach for aligning mass spectrometry data based on Gaussian scale space analysis has been presented [12]; however, statistical approaches rely upon having a sufficient number of experiments to estimate a data model. By incorporating the both clique finding and local optimization, our method can leverage larger numbers of experiments but can also function with smaller numbers.

Our method is also similar in spirit to certain clustering algorithms [13]; however, these algorithms are not specialized for data alignment. Some require prior knowledge of the total number of objects, which is not available to us. Others require various other parameters whose selection is less obvious than the error bounds derived from the sensors. Many clustering algorithms are also based on greedy local decisions to merge or divide groups of features, rather than more general optimization; such algorithms have been applied to the spectrum alignment problem in prior work [14][15]. We have also experimented with the CAST clustering algorithm [16] (results not shown), but found the performance was much slower, and the required parameters and similarity function were more difficult to select.

Our method does not address the related problem of *calibration*, in which transformations are applied to each data set to compensate for changes in the experimental equipment and conditions. In practice, we apply a combination of local translations by phase correlation [17] and dynamic time warping [18] to each data set to perform a "star" alignment to a chosen reference. From the perspective of our algorithm this is not theoretically necessary, but in practice allows us to reduce the size of the error bounds thereby improve the efficiency and quality of our result.

Aknowledgements

We wish to thank Vineet Bafna, Shibu Yooseph, and Knut Reinert for helpful discussions in the formative stages of the project. We aknowledge Albert Shieh for providing the methods for calibration. Finally, we thank Lap Ho at the Mount Sinai School of

Medicine for providing the real data which we used to generate our results for this paper.

References

1. Petricoin, E., Liotta, L.: Mass spectrometry-based diagnostics: the upcoming revolution in disease detection. Clin. Chem. 49, 533–534 (2003)
2. Diamandis, E.P.: Mass spectrometry as a diagnostic and a cancer biomarker discovery tool. Molecular and Cellular Proteomics 3.4, 367–378 (2004)
3. Aebersold, R., Mann, M.: Mass spectrometry-based proteomics. Nature 422, 198–208 (2003)
4. Semmes, O., et al.: Evaluation of serum protein profiling by surface-enhanced laser desorption/ionization time-of-flight mass spectrometry for the detection of prostate cancer: I. assessment of platform reproducibility. Clin. Chem. 51, 102–111 (2005)
5. Baggerly, K., et al.: Reproducibility of seldi-tof protein patterns in serum comparing data sets from different experiments. Bioinformatics 20, 777–785 (2003)
6. Turnbull, B.W.: The empirical distribution function with arbitrarily grouped, censored, and truncated data. Journal of the Royal Statistical Association, Series B. 38, 290–295 (1976)
7. Maathuis, M.: Reduction algorithm for the NPMLE for the distribution function of bivariate interval censored data. Journal of Computational and Graphical Statistics 14, 352–362 (2005)
8. Kirkpatrick, S., Gelatt, C., Vecchi, M.: Optimization by simulated annealing. Science 220, 671–680 (1983)
9. Bogaerts, K., Lesaffre, E.: A new fast algorithm to find the regions of possible mass support for bivariate interval censored data. Technical Report 0312, IAP Statistics Network (2003)
10. Song, S.: Estimation with Bivariate Interval Censored Data. PhD thesis, University of Washington (2001)
11. Gentleman, R., Vandal, A.C.: Computational algorithms for censored data problems using intersection graphs. Journal of Computational and Graphical Statistics, pp. 403–421 (2001)
12. Uy, W., Li, X., Liu, J., Wu, B., Williams, K.R., Zhao, H.: Multiple peak alignment in sequential data analysis: A scale-space approach. IEEE/ACM Transactions on Computational Biology and Bioinformatics 3(3), 208–219 (2006)
13. Duda, R.O., Hart, P.E., Stork, D.G.: Pattern Classification, 2nd edn. John Wiley & Sons, Inc, New York (2001)
14. Bellew, M., Coram, M., Fitzgibbon, M., Igra, M., Randolph, T., Wang, P., May, D., Eng, J., Fang, R., Lin, C.W., Chen, J., Goodlett, D., Whiteaker, J., Paulovich, A., McIntosh, M.: A suite of algorithms for the comprehensive analysis of complex protein mixtures using high-resolution lc-ms. Bioinformatics 22(15), 1902–1909 (2006)
15. jun Li, X., Yi, E.C., Kemp, C., Zhang, H., Aebersold, R.: A software suite for the generation and comparison of peptide arrays from sets of data collected by liquid chromatography-mass spectrometry. Molecular & Cellular Proteomics 4(9), 1328–1340 (2005)
16. Ben-Dor, A., Yakhini, Z.: Clustering gene expression patterns. Journal of Computational Biology 6(3/4), 281–297 (1999)
17. Kuglin, C.D., Hines, D.C.: The phase correlation image alignment method. In: Proceedings of the IEEE International Conference on Cybernetics and Society (1975)
18. Salvador, S., Chan, P.: Fastdwt: toward accurate dynamic time warping in linear time and space. In: Proceedngs of the KDD Workshop on Mining Temporal and Sequential Data (2004)

Author Index

Lecture Notes in Bioinformatics

Vol. 3082: V. Danos, V. Schachter (Eds.), Computational Methods in Systems Biology. IX, 280 pages. 2005.

Vol. 2994: E. Rahm (Ed.), Data Integration in the Life Sciences. X, 221 pages. 2004.

Vol. 2983: S. Istrail, M.S. Waterman, A. Clark (Eds.), Computational Methods for SNPs and Haplotype Inference. IX, 153 pages. 2004.

Vol. 2812: G. Benson, R.D.M. Page (Eds.), Algorithms in Bioinformatics. X, 528 pages. 2003.

Vol. 2666: C. Guerra, S. Istrail (Eds.), Mathematical Methods for Protein Structure Analysis and Design. XI, 157 pages. 2003.